PLANT INVASIONS: STUDIES FROM NORTH AMERICA AND EUROPE

Edited by J.H. Brock, M. Wade, P. Pyšek and D. Green

Backhuys Publishers, Leiden, 1997

ISBN 90-73348-23-4

Printed in the Netherlands

TABLE OF CONTENTS

PREFACE

Alien invasive plants are becoming recognized by some authorities as another component of global environmental change. When an alien plant(s) overtake the native vegetation and essentially develop a monoculture, it can be said that the environment of that area has changed. The decrease in biological diversity is a concern among many ecologists and invasive plants are seen as a threat to ecosystem health. In studying invasion ecology, an introduced species should be treated as an alien to the natural ecosystem because it would not be there without human intervention.

Alien invasive plants have become firmly naturalized in most habitats of North America and the world. Management attempts that extol eradication of these alien species most likely will be futile. Management to control plants on critical portions of the habitat can be achieved with economic, social, and perhaps political costs. Many times the ecological and economic costs are bearable, but the social and political concerns curtail widespread successful vegetation management programs. Existing vegetation management tools will effectively control alien invasive plants if there is both the political and social concensus for vegetation management.

This book continues the compilation of information about alien invasive plants, and especially what characteristics of the plants can be used to provide an early warning system about their invasive potential. This book builds on workshop materials presented at an October, 1995 conference at Arizona State University, Tempe, Arizona, USA, and complements papers in **Plant Invasions: General Aspects and Special Problems**, published in 1995 by SPB Academic Publishing, and **Ecology and Management of Invasive Riverside Plants** published in 1994 by Wiley & Sons.

I wish to thank Max Wade of Loughborough University (United Kingdom) Petr Pyšek, Institute of Botany (Czech Republic) and Douglas Green, Arizona State University (United States) for their help in reviewing the papers contained in this book. I also thank the authors of the papers for their research and creative activities.

John H. Brock

PREFACE

Alien invasive plants are becoming recognized by some authorities as another component of global environmental change. When alien plant(s) overtake the native vegetation [illegible] consequent [illegible] of that [illegible] decrease in biological diversity is a concern among many ecologists and [illegible] threat to ecosystem health. In studying invasion ecology [illegible] could be treated as an alien in the natural ecosystem because it would not be there without human intervention.

[illegible]

[illegible] and the [illegible] of the [illegible] can [illegible] provide an early warning [illegible] about [illegible] potential. This book [illegible] workshop [illegible] Arizona State [illegible] Tempe, Arizona, USA, and complements papers in Plant Invasions: General Aspects and Special Problems, published [illegible] Ecology and Management of Invasive Riverside Plants [illegible]

[illegible]

PREDICTING PLANT INVASIONS: MAKING A START

Max Wade
International Centre of Landscape Ecology, Department of Geography
Loughborough University, Loughborough, Leicestershire LE11 3TU, United Kingdom

Abstract

The alien species found in the flora of a country or region are likely to contain a number of potentially invasive plants, candidate species, which at some time in the future will present serious problems to human activities. The prediction of such invasive species is considered in relation to the growing knowledge of the ecology of plant invasions. The pattern of such an invasion is broken down into four stages. The critical phase for prediction is stage I, the lag phase, in association with the factors which trigger a species to go from stage I through stage II to stages III and IV, the invasive phase. These critical factors include disturbance and climatic change. Overall, there is a wide range of factors and generalisations that cannot be made either in terms of factors or of species. The existing infrastructure underpinning both the surveillance of alien plant species and the institutional frameworks and attitudes of both governmental and non-governmental organisations are considered for the UK and the USA. It is concluded that whilst there is still much to learn about the process of invasion, the lag phase usually lasts a number of years and hence there is scope to identify potentially invasive species and take appropriate action to prevent them moving into stages III and IV.

Introduction

"Nowadays we live in a very explosive world, and while we may not know where or when the next outburst will be, we might hope to find ways of stopping it or at any rate dampening down its force". So began Charles Elton in his book on the ecology of plant and animal invasions (Elton 1958). Whilst there has been considerable research into the "dampening down" of plant invasions (or "outbursts") (e.g. Drake *et al.* 1989; Macdonald *et al.* 1986; de Waal *et al.* 1994; Cronk and Fuller 1995), little progress has been made towards preventing such invasions occurring in the first place. This is a surprising situation given the scale of the problems caused by alien species around the world (Cronk and Fuller 1995; Humphries *et al.* 1991; Macdonald *et al.* 1986; U.S. Congress, Office of Technology Assessment 1993) and the research effort which has been directed at understanding the ecology of those plants which have become invasive (e.g. Groves and Burdon 1986; Drake *et al.* 1989; di Castri *et al.* 1990; Pysek *et al.* 1995). The enormity of the problem is not in dispute. Recent estimates of the number of species of alien vascular plants in different individual countries of the world typically exceed a thousand: 1,360 aliens in the British Isles (ca. 40% of the total flora) (Ellis 1994); 1,500–2,000 species introduced to Australia since European settlement, over 200 of which are considered to be noxious weeds (Humphries *et al.* 1991; Parsons and Cuthbertson 1992); and over

Plant Invasions: Studies from North America and Europe, pp. 1-18
edited by J.H. Brock, M. Wade, P. Pyšek and D. Green

2,000 species of non-native plants in the United States, many causing significant economic and ecological damage (U.S. Congress, Office of Technology Assessment 1993). The movement of plant species into a region due to human activity is a continuous process and the number of species is significant. At a national level, the Botanical Society of the British Isles (BSBI) reports on 50–100 alien species annually in its serial "BSBI News" (e.g. Ellis 1995). At the more local level, Pyšek and Mandák (1997) describe the changes in numbers of alien species in settlements in the Czech Republic.

Only a few alien species in a region will become invasive. Williamson (1993) estimated for the British flora that 10% of introduced species become established and 10% of these established species go on to become pests. This relies on the subjective notion of a pest and makes actual predictions difficult. Kowarik (1995) calculated that on the basis of woody species introduced into Brandenburg and Berlin, Germany (3,150 species), 10% began to spread, 2% become established and half of these became members of the natural vegetation, a 10:2:1 ratio which was found to be the same as that for the Central European vascular plant flora (12,000 species) (Kowarik 1995). Weeda (1987) working on numbers of neophytes and alien species in the Netherlands estimated that of approximately 7,000 species, 75 penetrated the natural vegetation, i.e. 1%, lending support to Kowarik's latter percentage. These calculations give a good indication of the number of candidate species which would need to be considered in screening for potentially invasive plants.

Those species which do become invasive exhibit a broad spectrum of life forms. Within aquatic and riparian habitats alone, there are submerged, floating, emergent and tree species, e.g. *Eichhornia crassipes* (e.g. Gopal 1987); *Lagarosiphon major* (e.g. Howard-Williams and Davies 1988); *Mimosa pigra* (e.g. Lonsdale *et al.* 1989) and *Melaleuca quinquenervia* (e.g. Myers 1983) respectively. These alien species have distributed themselves around the world causing environmental problems of international dimensions. Other habitats have been invaded with an equally diverse range of species and life forms (Cronk and Fuller 1995). The extension of distribution is a dynamic process, for example, *Fallopia japonica,* initially a problem of growing dimensions in Europe has now become invasive and problematic in certain parts of the USA (Brock *et al.* 1995). Not all alien invasive plants generate such chronic problems and some have desirable properties, for example, the flowers of *Buddleia davidii* are both attractive and an excellent source of nectar for butterflies and other insects (Owen and Whiteway 1980) creating a different attitude to this species. Alien plants in general contribute to the biodiversity of a region as evidenced by the figures quoted above and can be an interesting and worthwhile component of the flora (Hanson 1995; Barker 1996; Kendle 1996). The challenge of invasive plant management is to ensure that the environmental balance is not adversely disrupted.

There is a growing body of knowledge regarding the ecology of plant invasions and this paper focuses on the ability to predict those alien species which might become the invasive weeds of tomorrow disrupting the environmental balance and causing problems to human activity which cost large sums of money to solve. Understanding the theoretical ecology of the invasion process is of limited value if the infrastructure is not available to generate the necessary data to inform the process and prevent a species from being introduced into a new region and, or becoming invasive. This depends upon institutional frameworks and attitudes involving both government and non-governmental agencies. The emerging infrastructure in two contrasting geographical regions, the UK and the USA, is examined. Achieving the ability to predict the alien invasive species of tomorrow and providing the basis to stop "the next outburst" would be a real step forward from Elton's observation in 1958.

The terminology used to describe these alien plants has been reviewed by Pyšek (1995) and it is his recommended terminology which is used in this paper, in which those introduced plants are referred to as alien species. The term invasive is used for an alien whose distribution and/or abundance in a region is increasing, i.e. can be considered as a successful alien. A species which is considered to have the potential to become invasive is termed a candidate species. Prior to this, the plant can be described as a listed species. It is important to differentiate between listed and candidate species. There are so many alien species, i.e. listed species, existing within a region that controlling or eradicating all of them is unrealistic and would damage the biodiversity resource. When action is taken against a candidate species it becomes a target species. The paper concentrates on aquatic and riparian alien species.

Developing ecological strategies

Such are the problems posed by alien invasive weeds that much of the research effort has been directed at exploring ways of containing their spread and reducing their distribution. This was initially conceived in terms of control, for example, by the application of herbicides, but has more recently been recognised as a plant management issue where an understanding of not only control measures but the ecology and biology of the target species is integral to successfully overcoming the problems (e.g. Kluge *et al.* 1986; Kluge and Erasmus 1991). Investigations have been undertaken into the ecological processes underpinning the development of a newly introduced plant species into a noxious weed. This has involved examining the history of such species; comparing their biology and ecology, and searching for those factors which have enabled them to become invasive given that many such species in their indigenous environment are benign, e.g. *Lythrum salicaria*, *Melaleuca quinquenervia* (Fairley and Moore 1989) and *Fallopia japonica* (Wade *et al.* 1996). In the case of the latter species, Figure 1 shows the changes in its distribution in the British Isles. The history of this spread has been reconstructed in detail by Conolly (1977) and, as with most such species there was a significant time-lag, in this instance of 50 years, before *F. japonica* began to develop into its invasive stage (see also Kowarik 1995). There is now a substantial body of information concerning this species (Beerling *et al.* 1994) which begs the question, would it have been possible to predict that *Fallopia japonica* would have become an invasive weed? Such an approach would have avoided environmental disruption and the substantial costs of managing a plant in the invasive phase, management often requiring the use of large amounts of herbicide (Stensones and Garnett 1994).

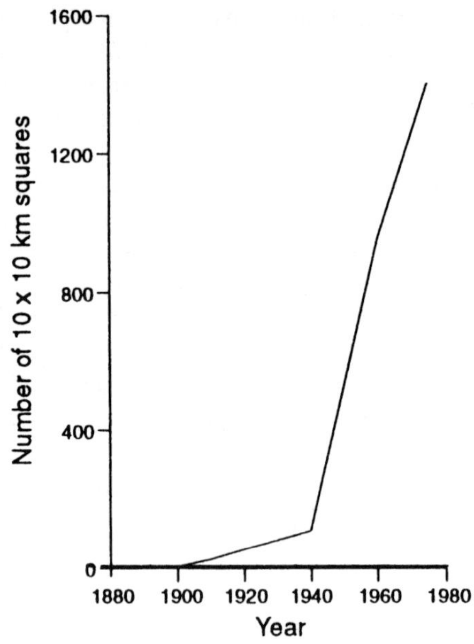

Fig. 1. Changing distribution of *Fallopia japonica* 1886 to 1976. (Source: Conolly 1997 and Beerling *et al.* 1994)

The pattern of invasion

An alien species such as *Fallopia japonica* has to pass through a number of phases before it reaches the invasive stage beyond which it could cause serious problems to humans. The pattern of spread has been reconstructed from historical data for a number of invasive aliens and has been found to be similar (Figure 2). The rates of establishment of alien species can be used to identify the different phases of invasion and these can be expressed as number of locations or recording units, e.g. 10 km^2 squares, colonised by the species per unit time, or distance travelled per unit time. Not all species which become established within the flora of the invaded region cause problems, e.g. *Mimulus guttatus* in the UK. The process can lead eventually to the naturalisation of the species (Kornas 1990).

Initially, a species has to be transported into a new region which in the case of aliens is usually supported deliberately or unwittingly by human activities. Legislation has been established in many countries to restrict this process (see Putting prediction into practice below). The next phase, stage I, is one of establishment in which the species persists at a site for a long time and may undergo reproduction (Figure 2). Many alien species are incapable of becoming established, in some cases with seedlings failing to develop into mature plants. Many species may remain sterile or if they flower they do not fruit, or the fruits are not viable. Of 799 introduced species found in and around Montpellier, France, 692 failed to become established (Thellung 1912) being termed ephemerophytes (Kornas 1990). This first phase is characterised not so much by the rate of establishment which is likely to be very low but the ability of the species to maintain itself at relatively few locations over a number of years. It

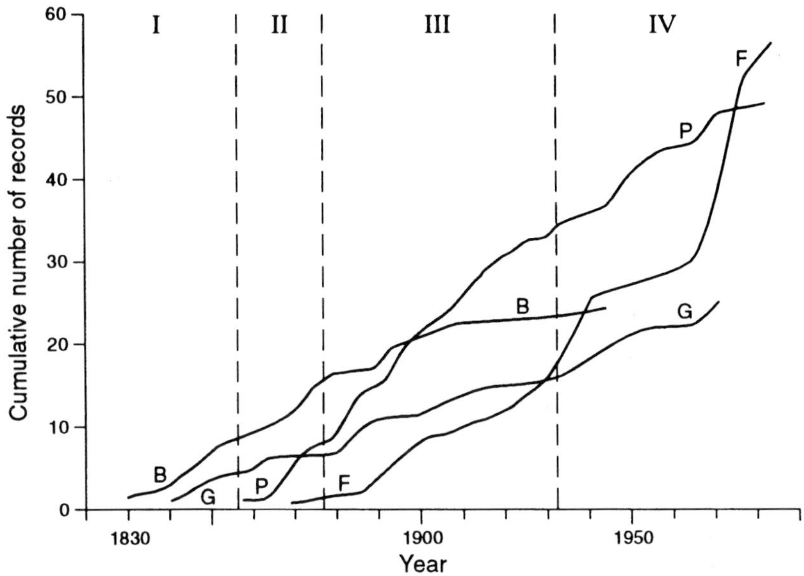

Fig. 2. Stages of invasion (I–IV) in relation to number of *Impatiens parviflora* records from different habitats in central Europe from 1830 to 1971. B = botanical gardens; G = gardens; P = parks and cemeteries; F = forests. Sources: Trepl (1984) and Kornas (1990)

is known as the lag phase (Kovarik 1995). It is during this lag phase that action should be taken to prevent a potentially invasive species from developing through stage II to stages III and IV. However, given that the species is likely to be known from only tens of sites or a handful of administrative areas, identification is likely to be very difficult. On the other hand, the species will have been evident for a number of years, even decades (Table 1), during which time appropriate investigations could have been undertaken to determine the potential of such species to become invasive.

In stage II, the alien species becomes permanently established in one or more sites which are typically strongly disturbed (Trepl 1984; Kornas 1990). In stage III, the immigrant begins to colonise sites which are only slightly disturbed and finally in stage IV, it succeeds in penetrating into wholly undisturbed sites (Figure 2). The second phase (stage II) sees an increase in the rate of colonisation typically with an extension of the distribution of the species throughout the region. If action had not been undertaken to manage the species in stage I, it is essential during this phase to contain its spread and development on into the invasive phase. The invasive phase (stages III and IV) is characterised by the rapid spread of the species throughout the region, the rate of which can be exponential. It is during this phase that the plant causes problems and incurs a growing expenditure due to disruption caused and in attempts to control it.

The invasive phase can last for a significant period of time. Some species remain in a permanently invasive mode, e.g. *Acer pseudoplatanus* in the UK whereas others undergo a decline in distribution and/or vigour and become assimilated into the

Table 1. A review of the lag times for a range of plant species

Crassula helmsii	UK	(Dawson and Warman 1987)	20 years
Fallopia japonica	UK	(Conolly 1977)	50 years
Impatiens glandulifera	Czech Republic	(Pyšek and Prach 1995)	40 years
Tree species (184 species)	Germany	(Kowarik 1995)	6% up to 50 years 25% up to 100 years 51% up to 200 years 14% up to 300 years 4% up to >300 years
Parthenium hysterophorus	Australia	(Auld *et al.* 1982/83)	20 years

indigenous flora, e.g. *Elodea canadensis* (Simpson 1984). Little is known as to what dictates the fate of a species after it has passed through its invasive phase.

Predicting invasiveness

Predicting the pattern of invasion in any detail will be a significant challenge. It will vary from species to species (Table 1) and for a given species from one region to another. Perrins *et al.* (1993) compared the invasion of the UK by *Impatiens glandulifera, I. parviflora* and *I. capensis* on the basis of the number of English vice-counties occupied (Figure 3), each species showing a different pattern. On the basis of these data, rates of spread were estimated, the maximum rates being 38, 24 and 13 km y^{-1} respectively over 170 years. The lag phase for *Impatiens glandulifera* in the Czech Republic was 40 years (first record in 1869) (Pyšek and Prach 1995) whereas in the UK it was almost 80 years (first record 1855) (Perrins *et al.* 1993). Pyšek and Prach (1995) found that, in contrast, the rates of invasion in the Czech Republic and the UK were similar. The regression coefficient b of the log-transformed cumulative number of localities over time was 0.053 for the Czech Republic and 0.051 for the UK. In Slovakia and Bavaria (Germany), however, the invasion started later, in the 1960s and 1970s respectively, and was more rapid after those dates (b = 0.111 and 0.131, respectively). In all four countries and probably across Europe extensive invasion by these species has been obvious in the last three decades regardless of the time of introduction.

Species which are in an apparent lag phase are, by definition, potentially invasive and this invasiveness should be explored by reviewing autecological information and, where necessary, collecting data on critical aspects of the plants, e.g. reproductive strategy and germination requirements. Of particular importance is the amplitude of ecological tolerance known for the species. It is through such data that predictions can be made of factors which might initiate the invasive phase: increase in disturbance, e.g. *Heracleum mantegazzianum* (Tiley *et al.* 1996), rise in temperature, e.g. *Fallopia japonica* (Beerling 1993), and species reaching a particular habitat type, e.g. *Elodea canadensis* and the canal system (Sculthorpe 1967).

Predicting whether or not a species should be regarded as a candidate species is fraught with difficulties. This is well illustrated by consideration of species of *Impatiens* alien to Europe. This genus provides some useful examples of the different

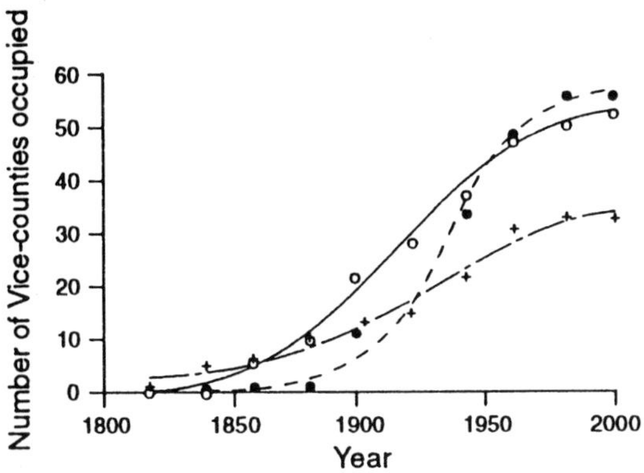

Fig. 3. Number of English vice-counties occupied by *Impatiens glandulifera* (●), *I. capensis* (+) and *I. parviflora* (o). The points are the actual data and the lines are fitted logistic curves (Source: Perrins *et al.* 1993)

stages of invasiveness which a species can reach and, given the pest status of certain *Impatiens* species in other parts of Europe, e.g. *I. glandulifera, I. parviflora* and *I. balfouri*, there should have been good reason for concern that other alien species in this genus might have become invasive in the UK. The species of *Impatiens* have similar growth forms and origins, e.g. *I. glandulifera* and *I. balfouri* are both from the Himalaya region (Williamson 1996). Likewise seed dispersal mechanisms and seed characteristics are ostensibly similar (Trewick and Wade 1986). On closer inspection, these species are found to behave quite differently.

I. balfouri, introduced from the Himalayas, was found by the river Thames in London in 1979 but did not persist and has not been recorded since (Burton 1983; Clement and Foster, 1994). This plant did not reach stage II (Figure 2) and appears to be unsuitable to conditions in the UK. *I. balfouri* is, nevertheless a problem alien species in the riparian habitat elsewhere in Europe and should remain under surveillance in relation to climate change. Two other species of similar status are *I. sultani* (probably a derivative from hybrids between the tropical *I. walleriana* Hook. f. and other species) and *I. balsamina* (from S.E. Asia). Both are casual garden escapes neither of which has established itself outside of cultivation (Burton 1983; Clement and Foster 1994) and neither of which would have been considered as candidate species based on their inability to either distribute to more than one or two locations nor to sustain a population at any of these locations.

I. capensis, an established garden escape from eastern North America has succeeded in passing through stages I, II and III having become naturalised on river banks especially in southern England in the late nineteenth century (Allen 1954), sometimes growing abundantly, and still increasing its distribution (Perrins *et al.* 1993). This plant would have been classed as a candidate species although it has not yet managed to establish itself in truly natural communities (stage IV).

Two species which have achieved this stage IV situation are *I. parviflora* and *I. glandulifera. I. parviflora* deserves special mention having spread all over Europe, initially as a typical ruderal but in the late 1800s moving into woods becoming firmly established in natural stands of deciduous forests, a remarkable situation in that few

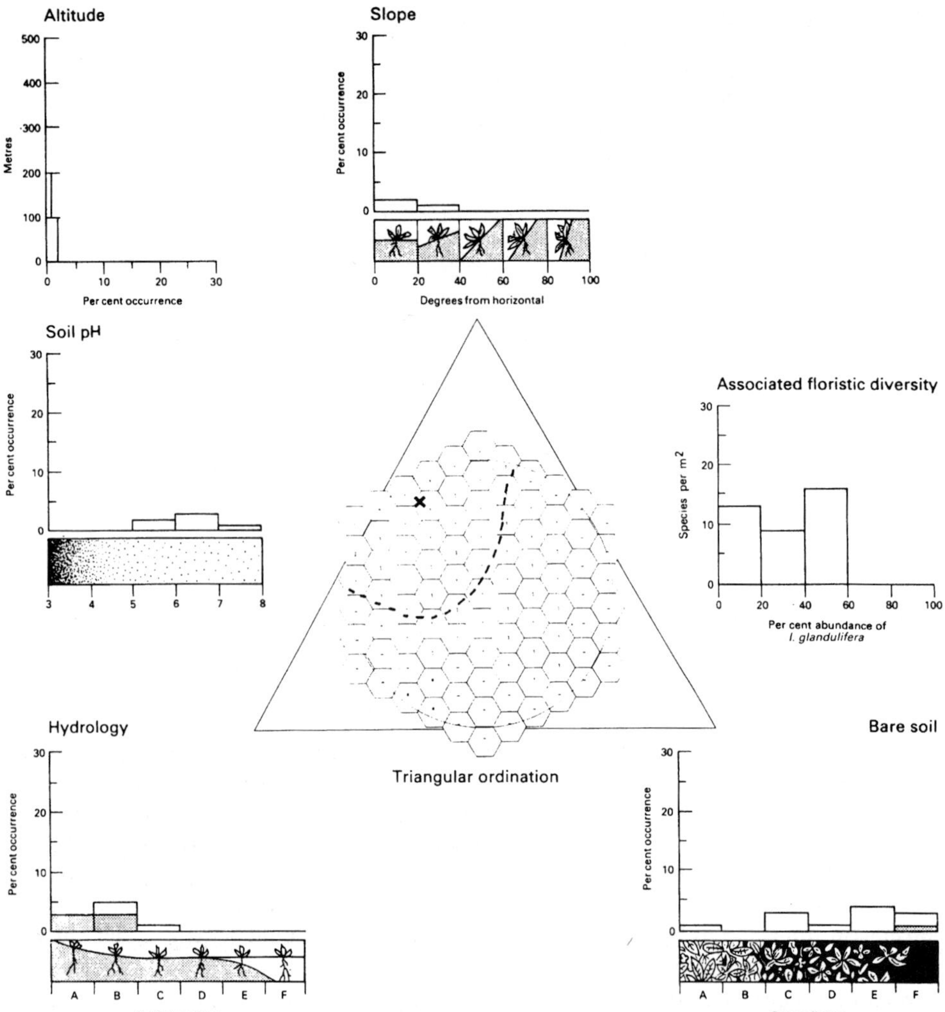

Fig. 4. Extract of autoecological account of *Impatiens glandulifera* Royle. (The Triangular ordination represents the relative constancy of the species in a matrix of vegetation types classified according to the strategies of the component species. Numbers in the cells indicate the percentage of samples in which the species occurs: – = absent; 1 = 1–10%; 2 = 11–20%; 3 = 21–30% etc., to 9 = 80% or above. The X indicates the approximate position of the maximum percentage occurrence of the species and a dashed contour delimits the region within which it achieves at least one-quarter of its maximum percentage occurrence.) Source: Grime *et al.* (1988)

aliens have truly succeeded in establishing themselves in entirely natural habitats (Kornas 1990) (Figure 2). In the UK it is thought to have been introduced from central Asia in Russian timber and is now naturalised in shady places in scattered localities especially in south-east England (Perrins *et al.* 1993). *I. glandulifera* is probably the most notable and certainly the most obvious alien *Impatiens* species in

Table 2. Comparison of three invasive alien riparian weeds

Fallopia japonica	*Heracleum mantegazzianum*	*Impatiens glandulifera*
Perennial	3 year cycle	Annual
Small white flowers in clusters	Small white flowers in enormous umbels	Large purple-pink flowers
Flowers from late August until September	Flowers from June until August	Flowers from July until October
No pollination (no male plants)	Pollination by insects	Pollination by insects
No seeds produced	Large numbers of seeds set in July to September	Large numbers of seeds set in August to October
No seeds	Seeds float	Seeds float and then sink
No seeds	Seeds require chilling	Seeds require chilling
No seed bank	Large seed bank of uncertain duration (2-14(?) years)	No known seed bank
Spreads via rhizome fragments	Perennates as root - no evidence of spread by roots	Does not spread via roots
Cut stems root at nodes under certain conditions	Cut stems do not root at nodes	Cut stems do root at nodes

the UK. The first records of naturalisation were from Middlesex and Hertfordshire in 1855 (Britten 1900) and this species has now established itself in relatively natural communities such as the *Epilobium* sub-community of *Phragmites australis-Urtica dioica*, and woodlands, e.g. *Alnus–Fraxinus–Lysimachia* woodland (Beerling and Perrins 1993). Elsewhere in Europe it has reached similar infestation levels as a riparian weed locally suppressing and endangering native species (Trautmann and Lohmeyer 1975) and even obstructing the rejuvenation of woody species (Lhotská and Kopecký 1966).

These examples indicate that because one species is invasive, this does not indicate that other species in the same genus are necessarily also likely to behave in a similar manner.

Ecological amplitude and autecology

The final phase of predicting invasiveness is the scoping of potential triggers in relation to existing knowledge of the candidate species and other closely related species. The scoping might confirm that under certain circumstances the species would become invasive. Alternatively it might identify aspects of the species or the environments/habitats it might invade which need investigation. The possible change in the climate associated with global warming and the greenhouse effect may not only trigger an increase in the rate of spread of already established although presently quiescent aliens, but may also allow new aliens to become established. Establishing the ecological amplitude for a candidate species needs to include available data on the species from its native region and, if it is known, from other regions. Use of these data alone could save significant environmental damage and economic cost. For example, good information is available on the invasiveness of a number of aquatic species (e.g. *Pistia stratiotes* and *Eichhornia crassipes*) yet it is true that when such a species is noted for the first time in a region little or no effort is made to deal with the plant at that early stage (Barker *et al.* 1996). There is a growing need for greater liaison between countries which are attempting to predict invasiveness and those in which the candidate species is native or already a problem weed. Sukopp and Starfinger (1995) draw attention to the relatively few examples of the comparison of a species' ecology in its native and synanthropic ranges. For example, despite *Fallopia japonica* becoming a problem in European countries in the 1950s and 1960s, it was not until the late 1980s that collaboration was established between European and Japanese botanists (Sukopp and Sukopp 1988) and in the UK even later (Wade *et al.* 1996).

Texts such as Drake *et al.* (1989) review the important features which will dictate the ecological amplitude of a given species, including pollination, regeneration, and seed and propagule dispersal. These must be studied in detail given that each species is likely to be unique. The reproductive strategy of an invasive species is critical to its success in establishing in, what, to that species, is an alien community. Could a superior reproductive strategy be the main reason why species such as *Heracleum mantegazzianum, Impatiens glandulifera*, and *Fallopia japonica* have achieved dominance in the riparian habitat? Table 2 compares certain reproductive characteristics of these three species revealing that they are very different. *H. mantegazzianum* and *I. glandulifera* both rely on insect pollination, whereas no male flowers are known from *F. japonica* in Europe and no viable seed is set. It is interesting to note that despite being alien species which are considered to have an impoverished invertebrate fauna neither *H. mantegazzianum* nor *I. glandulifera* has any apparent problem in achieving a substantial seed set (Beerling *et al.* 1994; Tiley and Philp 1997). The three species are very different in a number of respects and despite careful investigation no characteristic or suite of characteristics has been identified which explains their considerable success in the riparian habitat (de Waal *et al.* 1995).

A valuable template for collating and checking for completeness of a species ecological amplitude is that of the Biological Flora of the British Isles (e.g. Beerling and Perrins 1993; Beerling *et al.* 1994; Tiley *et al.* 1996). Grime *et al.* (1988) also provide an excellent basis for summarising the autecology of a species (Figure 4). Recent investigations of species such as *Fallopia japonica* have revealed the importance of understanding the genetics of the taxon and its relation to other species and hybrids. Such data, apart from chromosome number, are not usually included in autecological reviews and summaries, but could be very important in understanding the complexity of invasion or minimally in providing baseline data for future reference to determine whether a species has undergone adaptation and hence genetic change. Hybridisation between alien and native species has given rise to new taxa such as the fertile and spreading *Fallopia* × *bohemica* (Bailey *et al.* 1996). Backcrossing has produced hybrid swarms, e.g. *Calystegia sepium* × *sylvatica*, and in the case of *Centaurea jacea* × *nigra*, an alteration in the genetic make-up of the native species, even though the alien species had died out (Clement and Foster 1994).

Putting prediction into practice

Comprehensive and accurate recording of alien plants as they are introduced into a region is the starting point for predicting an invasive plant and it is fortunate that alien plants have long held a fascination for botanists and that considerable effort has been expended in the recording of their distribution (e.g. Dunn 1905; Clement and Foster 1994). Surveillance must be carried out throughout the region under consideration including both rural and urban areas, the latter often serving as the source of introduction into the "wild" (e.g. *Fallopia japonica* (Conolly 1977). More demanding is the identification of alien species which will not be described in the flora of the region and, hence, be relatively unknown. Some taxa will present particular problems (e.g. *Taraxacum*) and hybridisation creates further difficulties (e.g. *Fallopia*-Bailey *et al.* 1996). Some regions will be well recorded, others under-recorded, in the case of the British Isles, Ireland (Clement and Foster 1994).

Whilst continuous surveillance is important, the lag phase for invasive species is such that the regularity of sampling can be measured in terms of years rather than months and effort should be expended accordingly. Such surveillance has been developed in recent years by the Botanical Society of the British Isles (BSBI) which has established an Alien Study Group inaugurated in 1993, the main aim of which is to study those aliens established in Britain and Ireland that appear to be increasing their rate of spread into the countryside and could pose a threat to native species. Aliens already widely established such as *Fallopia japonica* and *Epilobium ciliatum,* while not excluded from the list would have lower priority (Ellis 1994). The Alien Study Group will also: a. compile a register of aliens and seek volunteers to adopt a species or group of species and monitor their spread; b. publish a regular newsletter to keep group members informed as to what was going on and provide a medium for interchange of ideas with a summary being published in the society's "BSBI News" so that the whole membership would be informed of progress and invited to contribute ideas; c. devise simple experiments to study the rate of increase or competitive ability of aliens compared with native species; d. compile lists of books and other publications including sources of illustrations which are useful for identifying alien plants; and e. prepare guides to identification of groups of aliens where there was thought to be a need. Alien species will be included in the new national mapping scheme for the British Isles, "Atlas 2000" which will further increase awareness of these species, and allow their accurate identification, often for the first time (Dines 1996). In addition to noting newly introduced species, surveillance should also pay attention to species which have failed to establish. This information would be useful in describing general trends and in making decisions about such a species should it reappear in the future at another site, e.g. *Heracleum mantegazzianum* has disappeared from a significant number of sites in Northern Ireland (Wade *et al.* 1997) and a follow up study could identify the reasons for such losses.

The existence of past records for alien species in a region is of considerable value in reconstructing the invasion process, e.g. in determining the length of the lag phase (Perrins *et al.* 1993) and the direction and nature of spread (Sculthorpe 1967, Simpson 1984). Records can exist for some regions over as long as 100 years or so (e.g. Dunn 1905) and the alien flora of the British Isles has been thoroughly reviewed by Clements and Foster (1994) and Ryves *et al.* (1996).

A centralised recording scheme would collate and store these data and could identify species which persist at sites and extend their distribution according to predetermined criteria, e.g. present in more than two 10 km^2 recording squares over a five year period. This database would also enable those species which have failed to establish to be identified. This role is currently being undertaken by the Institute of Terrestrial Ecology's Biological Records Centre, and the Biological Records Centre could interact with the database to identify such trends as those species increasing in their distribution.

Surveillance should not necessarily be restricted to a given region or nation. Useful indicators and confirmation of concern regarding a species can be achieved by observing the behaviour of the species in other proximal geographical areas. For example, *Impatiens glandulifera* was extending its distribution during the 1970s and 1980s in Croatia (Lukac 1989); Italy (Costalonga 1987); North Ossetian (the Soviet Union) (Komzha and Popov 1990); Slovakia (Bohušová 1992); Belgium and Luxembourg (Bruynseels *et al.* 1986); and Germany (Koenies and Glavac 1979).

The next stage in the process of predicting potentially invasive species is to establish a monitoring or co-ordinating centre which serves to link institutions and to provide advice and information on invasive species at a national level. This centre would be responsible for issuing management recommendations and could have a role in co-ordinating strategy development for individual species.

Examination of existing practice reveals that there are organisations which collect or hold data and are able to contribute towards the prediction of invasive species, and some have even gone this far (e.g. Dawson and Warman 1987) though none has attempted to take on the role of a co-ordinating centre. The efforts of different organisations remain unintegrated and the key elements in the ecology of the invasion process are not recognised by government institutions.

In the United States of America, a more institutionalised approach is taken in dealing with alien species. For example in the state of Arizona, the Plant Services Division of the Arizona State Department of Agriculture has established a Noxious Weed Detection Programme the main role of which is the inspection of commercial shipments and nursery stock for alien plants (E. Hall, pres. comm. 1995). This focuses on weed plants and other vegetative material, e.g. fragments of *Hydrilla* caught on the propeller blades of boats, and is enforced through the Arizona Noxious Weed Regulations (Arizona Department of Agriculture 1994a) which include regulations dealing with those alien species already present in Arizona and those excluded from the state. The Arizona Department of Agriculture also enforces the Federal (USDA) Noxious Weed Act of 1974, although the efficacy of this instrument has been called into question given that in 1993 the list of noxious species was only 93, ignoring at least 750 species including *Myriophyllum spicatum* and *Schinus terebinthifolia*. The Department is a member of the Arizona InterAgency Noxious Weed Committee established in April 1992, the other four members being the Department of the Interior's Bureau of Land Management and Bureau of Indian Affairs, the US Department of Agriculture's Animal Plant Health Inspection Service and the Arizona Department of Transport. This inter-agency committee initially produced a list of noxious weeds known to occur in Arizona which was followed up by a distribution mapping programme, the data from which will be managed through a database package, Paradox, maintained in the Plant Services Division. Other roles of the inter-agency committee are raising public awareness and co-operating with the University of Arizona and New Mexico State University to run noxious weed field schools dealing with such aspects as plant recognition and control techniques. The Plant Services Division has established contact with a non-governmental organisation, the Arizona Native Plants Society which has a growing interest in exotics (alien plant species). (There has been surveillance of agricultural weeds through Agriculture Watch since 1983).

Legislation of importance in the control of alien invasive species includes the Arizona Seed Law (Arizona Department of Agriculture 1994b) passed by the Arizona State Legislature. This, for example, specifies the number of weed seeds allowed per unit mass of commercial seeds, except for some which are not allowed at all, e.g. *Centaurea solstitialis*. This law also bans the sale of certain types of seeds, e.g. all species of genus *Ipomoea*. The Arizona revised statutes allow the enforcement of the state Noxious Weed Regulations which includes a state noxious weed list.

Legislation has an important role in reducing the number of alien species arriving in a region and, if successful this will reduce the problems in years to come. From the point of view of this paper tomorrow's invasive weeds are already present in any given region. They have probably been so for some time and we are probably familiar with them. Stein (1996) found that 20 of the worst weeds in the USA, including *Lythrum salicaria* and *Lonicera japonica* are currently being sold by nurseries in the USA and the National Association of Exotic Pest Plants noted that nurseries in Florida have sold 80% of the state's exotic pest plants (Tenenbaum 1996). The pattern of control and legislation varies from state to state, for example California, the Pacific North West, Florida and Tennessee have established Exotic Pest Plant Councils. The former (CalEPPC) is dedicated to finding solutions to problems caused by non-native pest plant invasions focusing on wildlands, i.e. natural areas that support native ecosystems. CalEPPC is a stateside organisation with an active membership including volunteers, land managers and scientists. It has drawn up a list of California's worst wildland weeds, developed a pest plant database available on the electronic network EcoNet and produced guidelines for the use of volunteers in exotic pest plant control projects. It held successful state-wide symposia in 1992 and 1993 and has co-sponsored regional workshops.

CalEPPC categorises the alien invasive species (exotic pest plants) into five groups: most invasive wildland plants, e.g. *Arundo donax* and *Elaeagnus angustifolia*; wildland pests of lesser invasiveness, e.g. *Ailanthus altissima*; red alert (species with potential to spread explosively; infestations currently restricted in size) e.g. *Lythrum salicaria*; species for which more information is needed, e.g. *Phyla nodifera*; and species being considered but not listed, e.g. *Centranthus ruber*.

Discussion

The logic of identifying a potentially invasive species during the first stage of its invasion in order to avoid environmental damage and save money is undeniable. Achieving accurate prediction however remains a challenge and at first glance might be considered unrealistic due to, not only the large number of alien species constituting the population from which a candidate species might emerge, but also the range of criteria which might trigger a species to pass from stage I of an invasion through stage II to stages III and IV. On the other hand, there is a considerable body of information accumulating upon which such a prediction process could be based. In the first instance, for some countries, there is an excellent database being amassed on the population of alien species, a database which includes details of first recordings and all known sites. The lag phase for many species which became invasive was remarkably long, in most cases measurable in decades. This period of time is valuable for instigating surveillance of those alien species which move into stage II (2–10% of alien species) and for extending the knowledge of invasions and their prediction. At the same time, strategy needs to be established and implemented at an institutional level for the management of a candidate species to prevent it from going from stage I to stage III. This is currently very much in its infancy and, in the UK for instance, there are no examples where a co-ordinated approach has been taken towards managing an invasive species outside of the agricultural environment even when it was well into stage III. Some countries are, however, moving in this direction (Andersen 1997). Improving on prediction needs to develop hand in hand with the reactive process for dealing with what could be tomorrow's weeds. Two examples from the UK illustrate this very well.

Crassula helmsii was first recorded in the UK in 1956 and since that time it has moved through stages I and II and into stage III. Dawson and Warman (1987) examined the spread of the species and extrapolated its distribution predicting its propensity to become a weed in both freshwater and estuarine habitats, a prediction supported by field observations in Australia, its country of origin. Over a number of years, as the species fulfilled the initial prediction, attention was drawn to the need to manage the invasion and protect habitats which were at risk (Dawson 1988, 1990, 1991). Whilst there was no disagreement that *C. helmsii* was becoming a problematic weed, no mechanism existed to control a species which occurred in a range of situations: ponds, drainage ditches, canals and reservoirs managed by a range of different institutions, and no concerted action was or has been taken. This example demonstrates that it is possible to identify a candidate species. The data needed to achieve this were initially based on records built up by a network of botanists across the UK through the BSBI. This was followed by sustained surveillance for the plant which had become a focus of interest for botanists and thorough autecological research and a definition of the species, ecological amplitude. Had an agreed institutional strategy existed to react to this information significant environmental damage and financial costs associated with this weed could have been avoided. This case study shows the plant moving from listed to candidate status and on into the invasive phase, and emphasises the need for background research and the maintenance of international and interstate networks.

Taking another example, *Lagarosiphon major*, this species was first recorded in the UK in 1944 and became a listed species. Despite its origins from sub-Saharan Africa, it has managed to not only sustain itself but to spread to a number of sites around the UK. Today, *L. major* should be considered as a candidate species based on its continued spread and the knowledge which has built up about its ecological amplitude, in this case not so much from its native Africa, but from New Zealand where it has become a serious weed in lakes and other aquatic habitats (Howard-Williams and Davies 1988). On the basis of this information, the prediction would be that the trigger for *L. major* to move into stages III and IV would be its distribution into lakes such as those found in the English Lake District and North Wales where the plant could displace natural isoetid plant communities already under stress from such factors as eutrophication, and interfere with water based activities, e.g. sailing and canoeing. The current requirements are a more detailed risk assessment including the potential costs associated with different degrees of invasion, further studies of the autecology of the plant particularly in relation to its mode of distribution, and the planning of a strategy, should there be a necessity for active management. Surveillance of the species' distribution needs to be sustained including monitoring of existing sites. There is often as much value in studying a site from which a species has disappeared as one in which it still remains. A strategy for managing *L. major* would need to involve not only those institutions and individuals on whose property the species already exists but those whose property might be adversely affected should the species find its way into their lakes and reservoirs. This would involve the Environment Agency (the statutory body which regulates land, air and water resources in England and Wales), English Nature and the Countryside Council for Wales (the statutory bodies which achieve, enable and promote nature conservation in England and Wales), at least two National Park authorities (Lake District and Snowdonia), and those scientists best able to advise on ecological aspects of the strategy. Items on the agenda for the first meeting between such institutions would be intensifying surveillance in areas most at risk from the species, raising public awareness and the provision of information, e.g. how to control the plant, preventing further spread of the species and undertaking control at existing sites in a strategic fashion. The value of establishing a Geographical Information System (GIS) to store and present data has been identified (Child and de Waal 1997) and such a GIS would provide an excellent basis for planning and implementing a strategy of management. A GIS could be of use in the decision making process associated with invasion management: a critical decision which needs to be taken is when should action be taken to react to the candidate species. The GIS could be used for determining this threshold which would not be based solely on number of sites colonised per unit time but, as in this case of *L. major*, could be linked into the species reaching certain sensitive locations, e.g. a certain distance from the Snowdonia National Park.

The infrastructure for reacting to a potential invasion does not exist at the present in the UK and no examples of such procedures have been found for other countries, most of the emphasis being placed on the prevention of the plants being brought into a particular country or state. Such legislation is important but consideration should also be given to the use of statutory instruments to prevent spread and require control within a country or state. This could apply to both *C. helmsii* and *L. major* in relation to the trade in aquatic plants. Both species originate from trade outlets being widely sold and used in garden and other ornamental ponds. Legislation either curbing or controlling the spread of these species would be a valuable way of slowing or stopping the spread of the plant. Local instruments such as bye-laws can be effective in raising awareness and instigating a more co-ordinated approach to control, for example *Heracleum mantegazzianum* in Edinburgh (Wade *et al.* 1997). In the case of *L. major* there is good information on the control of the species using herbicides and by cutting but this is not always the case, for example, *Crassula helmsii* poses serious problems in terms of its control (Dawson and Warman 1987). Research into susceptibility to different control measures should be included in the studies of ecological amplitude and autecology, and will involve both greenhouse and laboratory based studies. This paper has dealt solely with invasion by alien species which have the potential to become invasive in aquatic and riparian habitats. It is important to remember that it is not only alien species which have this propensity, indigenous species can also become weeds, e.g. *Urtica dioica* and *Typha latifolia*.

Acknowledgements

Everett L. Hall, Arizona Department of Agriculture kindly provided information on the management of alien plant species in Arizona.

References

Allen, H.G. 1954. The orange balsam. Journal of the Northamptonshire Natural History Society and Field Club 1954: 249-250.

Andersen, U.V. 1997. Monitoring invasive weeds at landscape level in Denmark. This volume.

Arizona Department of Agriculture. 1994a. Arizona Noxious Weed Regulations. Phoenix.

Arizona Department of Agriculture. 1994b. Arizona Seed Law (effective March 1994). Phoenix.

Auld, B.A., Hosking, J. and McFadyen, R.E. 1982/83. Analysis of the spread of tiger pear and parthenium weed in Australia. Australian Weeds 2: 56-60.

Bailey, J.P., Child, L.E. and Conolly, A.P. 1996. A survey of the distribution of *Fallopia* x *bohemica* (Chrtek and Chrtková). Watsonia 21: 187-198.

Barker, G. 1996. Alien and native – drawing the line. Ecos 17: 18-26.

Barker, P.J., Ferguson, C.M., Smout, I.K. and Wade, P.M. 1996. The management of weeds in irrigation and drainage channels: integrating ecological, engineering and economic considerations. Hydrobiologia 340: 163-172.

Beerling, D.J. 1993. The impact of temperature on the northern distribution limits of the introduced species *Fallopia japonica* and *Impatiens glandulifera* in north-west Europe. Journal of Biogeography 20: 45-53.

Beerling, D.J., Bailey, J.P. and Conolly, A.P. 1994. Biological Flora of the British Isles: *Fallopia japonica* (Houtt). Ronse Decraene. Journal of Ecology 82: 959-979.

Beerling, D.J. and Perrins, J.M. 1993 Biological Flora of the British Isles: *Impatiens glandulifera* Royle. Journal of Ecology 81: 367-382.

Bohušová, K. 1992. Rozšírenie druhu *Impatiens glandulifera* Royle na Slovensku. Bull. Slov. Bot. Spol. Bratislava 14: 7-15.

Britten, J. 1900. *Impatiens roylei* in England. Journal of Botany 38: 50-51.

Brock, J.H., Child, L.E., de Waal, L.C. and Wade, P.M. 1995. The invasive nature of *Fallopia japonica* is enhanced by vegetative regeneration from stem tissues. In: P. Pyšek, K. Prach, M. Rejmanek and P.M. Wade. (eds) Plant invasions: general aspects and special problems, pp. 131-139. SPB Academic Publishing, Amsterdam.

Bruynseels, G., Leten, M and Saintenoy-Simon, J. 1986. Complements á atlas de la flore Belge et Luxembourgeoise. IFBL Feuille de contact trimestrielle 4: 7-26.

Burton, R.M. 1983. Flora of the London area. London Natural History Society, Spottiswode Ballantyne, Colchester.

Child, L.E. and de Waal, L.C. 1997. Use of GIS in the management of *Fallopia japonica* in the urban environment. This volume.

Clement, E.J. and Foster, M.C. 1994. Alien plants of the British Isles. Botanical Society of the British Isles, London.

Conolly, A.P. 1977. The distribution and history in the British Isles of some alien species of *Polygonum* and *Reynoutria*. Watsonia 11: 291-311.

Cotalonga, S. 1987. 16. *Impatiens glandulifera* Royle (Balsaminaceae). In: Matrini, F. and Poldini, L. (eds) Segnalazioni floristiche dalla regione Friuli-Venezia Giulia. II. Gortania 9: 148-150.

Cronk, Q.C.B. and Fuller, J.L. 1995. Plant invaders. A People and Plants conservation manual. World Wide Fund for Nature. Chapman & Hall, London.

Dawson, F.H. 1988. The alien aquatic *Crassula helmsii* continues to expand its distribution in Britain. BSBI News (49), 42.

Dawson, F.H. 1990. The spread of *Crassula helmsii* (Kirk) Cockayne in Britain. Proceedings of conference on the biology and control of invasive plants, British Ecological Society Industrial Ecology Group, University of Wales, Cardiff, 20-21.

Dawson, F.H. 1991. An investigation of the control of *Crassula helmsii* by herbicidal chemicals (with interim guidelines on control – final report). Nature Conservancy Council, Peterborough.

Dawson, F.H. and Warman, E.A. 1987. *Crassula helmsii* (T. Kirk) Cockayne: is it an aggressive aquatic plant in Britain? Biological Conservation 42: 247-272.

de Waal, L.C., Child, L.E., Wade, P.M. and Brock, J.H. (eds) 1994. Ecology and management of invasive riverside plants. John Wiley, Chichester.

de Waal, L.C., Child, L.E. and Wade, P.M. 1995. The management of three invasive alien riparian plants: *Impatiens glandulifera* (Himalayan balsam), *Heracleum mantegazzianum* (giant hogweed) and *Fallopia japonica* (Japanese knotweed). In: Harper, D.M. and Ferguson, A.J.D. (eds) The ecological basis for river management, pp. 315-322. John Wiley, Chichester.

di Castri, F., Hansen, A.J. and Debussche, M. 1990. Biological invasions in Europe and the Mediterranean basin. Kluwer Academic Publishers, Dordrecht.

Dines, T. 1996. Atlas 2000 – the official launch. BSBI News (72): 3-5.

Drake, J.A., Mooney, H.A., di Castri, F., Groves, R.H., Kruger, F.J., Rejmanek, M. and Williamson, M. (eds) 1989. Biological invasions: a global perspective. John Wiley, Chichester.

Dunn, S.T. 1905. Alien flora of Britain. West, Newman & Co., London.

Ellis, R.G. 1994. Alien Study Group. BSBI News (65): 44-45.

Ellis, R.G. (ed.) 1995. BSBI News, (68) Botanical Society of the British Isles. 68 pp.

Elton, C.S. 1958. The ecology of invasions by animals and plants. Methuen, London.

Fairley, A. and Moore, P. 1989. Native plants of the Sydney District: an identification guide. Kangaroo Press, New South Wales.

Gopal, B. 1987. Water hyacinth. Aquatic Plant Studies 1, Elsevier, Amsterdam.

Grime, J.P., Hodgson, J.G. and Hunt, R. 1988. Comparative plant ecology. Unwin Hyman, London.

Groves, R.H. and Burdon, J.J. (eds) 1986. Ecology of biological invasions. Cambridge University Press, Cambridge.

Hanson, G. 1995. Bedfordshire wool aliens. BSBI News, (70), 36.

Howard-Williams, C. and Davies, J. 1988. The invasion of Lake Taupo by the submerged water weed *Lagarosiphon major* and its impact on the native flora. New Zealand Journal of Ecology 11: 13-19.

Humphries, S.E., Groves, R.H. and Mitchell, D.S. 1991. Plant invasions of Australian ecosystems. A status review and management directions. Australian National Parks and Wildlife Service, Canberra.

Kendle, A. 1996. Front line. The National Trust Magazine (77): 23.

Kluge, R.L. and Erasmus, 1991. An approach towards promoting progress with the control of woody alien invasive plants in Natal. South African Forestry Journal 157: 86-90.

Kluge, R.L., Zimmermann, H.G., Cilliers, C.J. and Harding, G.B. 1986. Integrated control for invasive alien weeds. In: Macdonald, I.A.W., Kruger, F.J. and Ferrar, A.A. (eds) The ecology and management of biological invasions in southern Africa, pp. 295–303. Oxford University Press, Cape Town.

Koenies, H. and Glavac, V. 1979. Über die Konkurrenzfähigkeit des Indischen Sprinkrautes (*Impatiens glandulifera* Royle) am Fuldaufer bei Kassel. Philippia 4: 47-59.

Komzha, A.L. and Popov, K.P. 1990. New data on the adventive flora of the North Ossetian ASSR. Botanicheskii Zhurnal (Leningrad) 75: 108-110 (In Russian).

Kornas, J. 1990. Plant invasions in Central Europe: historical and ecological perspectives. In: F. di Castri, A.J. Hansen and M. Debussche (eds) Biological invasions in Europe and the Mediterranean basin, pp. 19-36. Kluwer Academic Publishers, Dordrecht.

Kowarik, I. 1995. Time lags in biological invasions with regard to the success and failure of alien species. In: P. Pyšek, K. Prach, M. Rejmanek and P.M. Wade. (eds) Plant invasions: general aspects and special problems, pp. 15-38. SPB Academic Publishing, Amsterdam.

Lhotská, M. and Kopecký, K. 1966. Zur Verbreitungsbiologie und Phytozönologie von *Impatiens glandulifera* Royle an den Flussystemen der Svitava, Svratka und oberen Odra. Preslia 38: 376-385.

Lonsdale, W.M., Miller, I.L. and Forno, I.W. 1989. The biology of Australian weeds 20. *Mimosa pigra* L. Plant Protection Quarterly 4: 119-131.

Lukac, G. 1989. Nova nalazista vrste *Impatiens glandulifera* Royle u Hrvatskoj. Acta botanica Croatia 48: 165-168.

Macdonald, I.A.W., Kruger, F.J. and Ferrar, A.A. (eds) 1986. The ecology and management of biological invasions in southern Africa, pp. 295-303. Oxford University Press, Cape Town.

Myers, R.L. 1983. Site susceptibility to invasion by the exotic tree *Melaleuca quinquenervia* in southern Florida. Journal of Applied Ecology 20: 645-658.

Owen, D.F. and Whiteway, W.R. 1980. *Buddleia davidii* in Britain: history and development of associated fauna. Biological Conservation 17: 149-155.

Parsons, W.T. and Cuthbertson, E.G. 1992. Noxious weeds of Australia. Inkata Press, Melbourne.

Perrins, J., Fitter, A. and Williamson, M. 1993. Population biology and rates of invasion of three introduced *Impatiens* species in the British isles. Journal of Biogeography 20: 33-44.

Pyšek, P. 1995. On the terminology used in plant invasion studies. In: P. Pyšek, K. Prach, M. Rejmanek and P.M. Wade. (eds) Plant invasions: general aspects and special problems, pp. 71-81. SPB Academic Publishing, Amsterdam.

Pyšek, P. and Prach, K. 1995. Invasion dynamics of *Impatiens glandulifera* – a century of spreading reconstructed. Biological Conservation 74: 41-48.

Pyšek, P. and Mandák, B. 1997. Fifteen years of change in the representation of alien species in Czech village flora. This volume.

Pyšek, P. Prach, K., Rejmanek, M. and Wade, P.M. (eds) 1995. Plant invasions: general aspects and special problems. SPB Academic Publishing, Amsterdam.

Ryves, T.B., Clement, E.J. and Foster, M.C. 1996. Alien grasses of the British Isles. Botanical Society of the British Isles, London.

Sculthorpe, C.D. 1967. The biology of aquatic vascular plants. Edward Arnold, London.

Simpson, D.A. 1984. A short history of the introduction and spread of *Elodea* Michx in the British Isles. Watsonia 15: 1-9.

Stein, S. 1996. Planting Noah's garden. Houghton, New York.

Stensones, A. and Garnett, R.P. 1994. Controlling invasive weeds using glyphosate. In: de Waal, L.C., Child, L.E., Wade, P.M. and Brock, J.H. (eds) 1994. Ecology and management of invasive riverside plants, pp. 183-188. John Wiley, Chichester.

Sukopp, H. and Starfinger, U. 1995. *Reynoutria sachalinensis* in Europe and the Far East. In: P. Pyšek, K. Prach, M. Rejmanek, and P.M. Wade, (eds) Plant invasions: general aspects and special problems, pp. 151-159. SPB Academic Publishing, Amsterdam.

Sukopp, H. and Sukopp, U. 1988. *Reynoutria japonica* Houtt. in Japan und in Europa. Verhöffentlichungen Geobotanisches Institute. Eidgenoessische Technische Hochschule Stiftung Rübel in Zürich 98: 354-372.

Tenenbaum, D. 1996. Weeds from hell. Technology Review 99: 32-40.

Thellung, A. 1912. La flore adventice de Montpellier. Mitteilungen Botanisches Museum Universität Zürich 58: 1-728.

Tiley, G.E.D., Dodd, F.S. and Wade, P.M. 1996. Biological Flora of the British Isles: *Heracleum mantegazzianum* Sommier and Levier. Journal of Ecology 84: 297-319.

Tiley, G.E.D. and Philp, B. 1997. Observation of flowering and seed production in *Heracleum mantegazzianum* in relation to control. This volume.

Trautmann, W. and Lohmeyer, W. 1975. Zur Entwicklung von Rasenansaaten an Autobahnen. Natur und Landschaft 50: 45-48.

Trepl, L. 1984. Über *Impatiens parviflora* D.C. als Agriophyt in Mitteleuropa. Diss. Bot. 73: 1-400.

Trewick, S. and Wade, P.M. 1986. The distribution and dispersal of two alien species of *Impatiens*: waterway weeds in the British Isles. Proceedings EWRS/AAB 7th Symposium on Aquatic Weeds, 1986, pp. 351-356.

U.S. Congress, Office of Technology Assessment. 1993. Harmful non-indigenous species in the United States. OTA-F-565. U.S. Government Printing Office, Washington D.C.

Wade, P.M., Child, L.E. and Adachi, N. 1996. Japanese knotweed – a cultivated coloniser. Biological Sciences Review 8: 31-33.

Wade, P.M., Darby, E.J., Courtney, A.D. and Caffrey, J.M. 1997. *Heracleum mantegazzianum*: a problem for river managers in the Republic of Ireland and the United Kingdom. This volume.

Weeda, E.J. 1987. Invasions of vascular plants and mosses into the Netherlands. In: Joenje, W., Bakker, K. and Vlijm, L. (eds) Proceedings of the Koninklijke Nederlandse Akademie van Wetenschappen, Amsterdam. Series C: Biological and Medical Sciences. pp. 19-29.

Williamson, M. 1993. Invaders, weeds and the risk from genetically manipulated organisms. Experimentia 49: 219-224.

Williamson, M. 1996. Biological invasions. Population and Community Biology Series 15, Chapman & Hall, London.

ALIEN WOODY PLANTS IN A SONORAN DESERT URBAN RIPARIAN CORRIDOR: AN EARLY WARNING SYSTEM ABOUT INVASIVENESS?

John H. Brock and Michael C. Farkas
Environmental Resources, School of Planning and Landscape Architecture, Arizona State University, Tempe, AZ 85287-2005 USA

Abstract

A drainage channel constructed about decade ago for the purpose of routing runoff water into the Salt River, Phoenix, Arizona, has spontaneously developed a vegetation community. The water collects from a mixed watershed of agricultural, residential, and industrial land uses which discharges into the normally dry Salt River. The water flow is perennial with the greatest component being irrigation runoff. The area was sampled in late August and September, 1993 to document species composition and structure of the vegetation stands. The vegetation along the channel consists of trees, shrubs, and herbaceous species, many of which are alien to the native flora of the Sonoran Desert biogeographical region. Ten of the seventeen woody plants growing along the channel are alien species. The main source of the alien plants is from horticultural landscapes in this largely urbanized area. We believe this developing vegetative community may provide an early warning system for land managers as to which domestic species have invasive tendencies and potential to spread into other stream channels and occupy riparian habitats.

Introduction

Alien plants are those that have been transported by human activities from their native ranges into new ecosystems where they did not evolve. In some cases, some of the alien species we use for agricultural, ornamental, or recreational purposes invade the native flora. This invasion can threaten or eliminate native species, contribute to degraded environmental conditions, and degrade ecosystem processes (Vitousek 1986). Biological invasions are also believed to be a component of global environmental change (Vitousek *et al.* 1996). There is a tendency for invaded areas to have a more homogenized flora, and reduce ecosystem diversity. There is a desire among ecologists studying alien species to have the ability, or the knowledge base, from which a plant's potential invasive nature could be ascertained. In other words, which alien species will invade? The answer to this question however, has proven elusive. Papers contained in Pyšek *et al.* (eds) 1995, by Rejmanek, Kowarik, and Pyšek, Prach and Šmilauer all illustrate the complexity of predicting which introduced species may become invasive. Rejmanek (1995) reported that only very limited generalizations are available based on physiology, genetics, demography, and species behavior. Vitousek (1986) indicates that the presence or absence of nitrogen (N) fixing symbioses is important in that the N fixing species have advantages in a N limited community and can help explain some species' invasive nature. Kowarik

Plant Invasions: Studies from North America and Europe, pp. 19–35
edited by J.H. Brock, M. Wade, P. Pyšek and D. Green

(1995) presented information on time lags in species invasion and concluded that successful invasions were not only related to the plant's life history but also interactions with habitat modifications such as the degree of surface disturbance by humans, and the trend toward global warming. Additionally, based on his data set from Germany, Kowarik (1995) presented a 10:2:1 ratio for invasiveness. That is, of the < 10% all introduced species that begin to invade, only 2% of the introductions became firmly established, and only 1% of the introduced species invaded natural vegetation. Bazzaz (1986) related that while some species spread quickly, others require a long period of "coevolutionary" interactions for integration into the resident ecosystem, providing the time lag discussed earlier. The success of an invading species may rely on its phenotypic plasticity, ability to germinate immediately or have delayed germination, and variation in chromosome numbers (Pimentel 1986). Attempts have been made to model invasions (Roughgarden 1986) and the rate of spread of invading species increases in a linear fashion with time to the square root of the area occupied by the species. Pyšek *et al.* (1995) state that there is no simple biological indication of invasion success, and that intensive studies are needed of the potential invader and the target community. What alien species will invade requires a thorough study of the autecology of the species in question, successional pathways and synecology of the target community, and an inventory of the abiotic factors in which the community is sited. Even then, if the community being invaded is sufficiently variable, then predicting anything about an invasion has high variability (Roughgarden 1986). Although predicting the invasiveness of a species is difficult, using the principle of convergence of evidence should be applied to the process. As more research is reported from around the world, the patterns of what makes a species a successful invader will emerge.

The objective of this paper is to present information concerning plant species, especially the alien ones, that were documented to be spontaneously growing in a plant community along a constructed waterway. The project began as one to determine habitat quality, especially for water and sediments as influenced by vegetation (Farkas, 1994). The information collected from those studies, allows an opportunity to place several species on the potential invader list, which may serve as part of an early warning system. The community studied is a linear habitat which provides less resistance in the spread of species (Pyšek and Prach 1994). The habitat allows for water transport of seeds and propagules, provides moist sites for seedling germination, and some site disturbance from flood events which provides openings for establishment. Furthermore, this linear habitat connects to a major river corridor which could serve as a conduit to move invasive species from the urban setting to downstream rural areas.

Methods

Study site description

Phoenix lies at the northeastern edge of the Sonoran Desert in south-central Arizona (Fig. 1) at an average elevation of 330 m. The vegetation is designated as middle Sonoran Desert Shrub (40-2AZ) by the Natural Resouces Conservation Service (1995). The average annual temperature is 22.5° C, with January and July average temperatures being 12.0° C and 34.2° C, respectively. Precipitation usually conforms

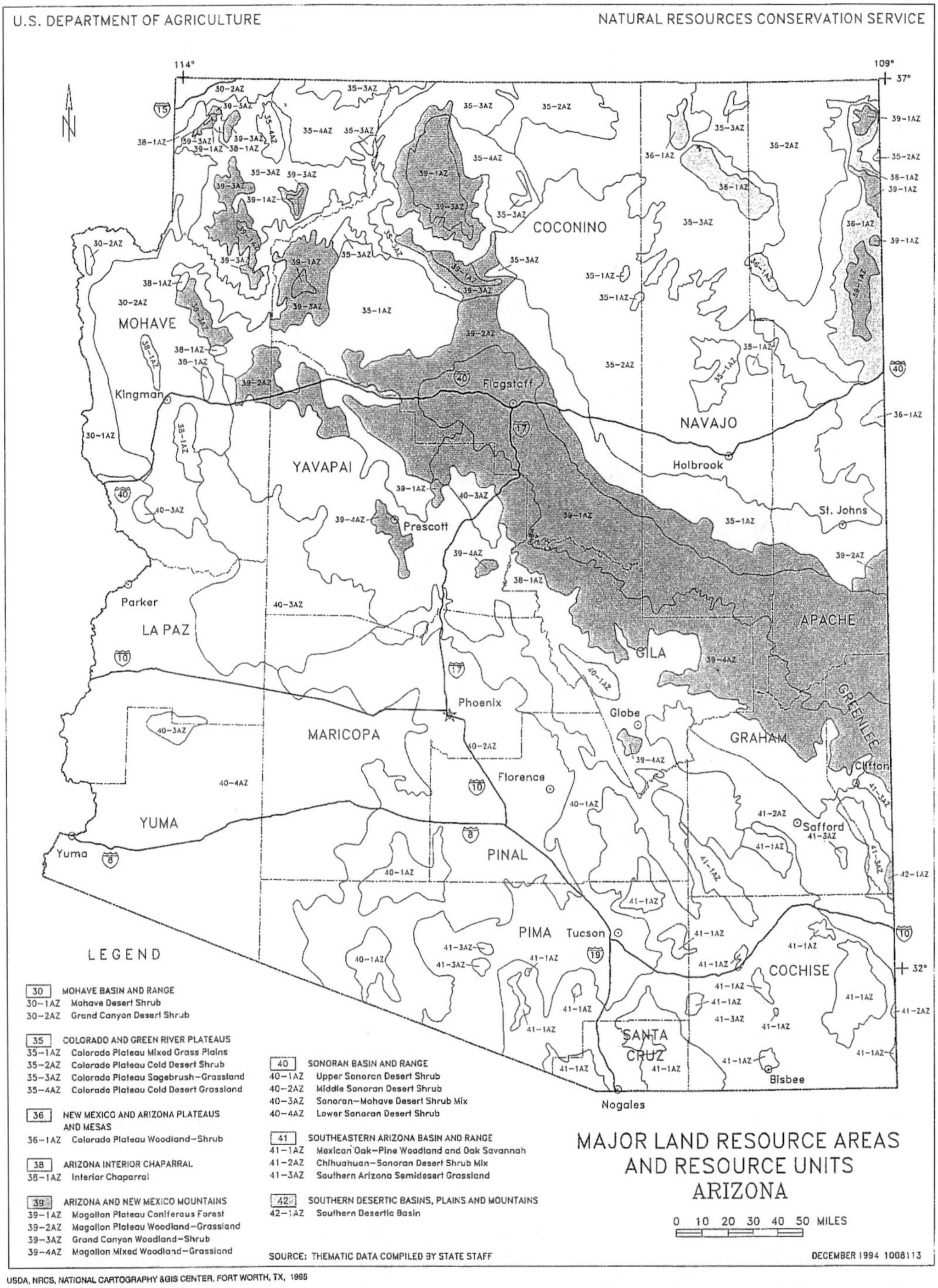

Fig. 1. Map of major land resources and the location of Phoenix in south-central Arizona.

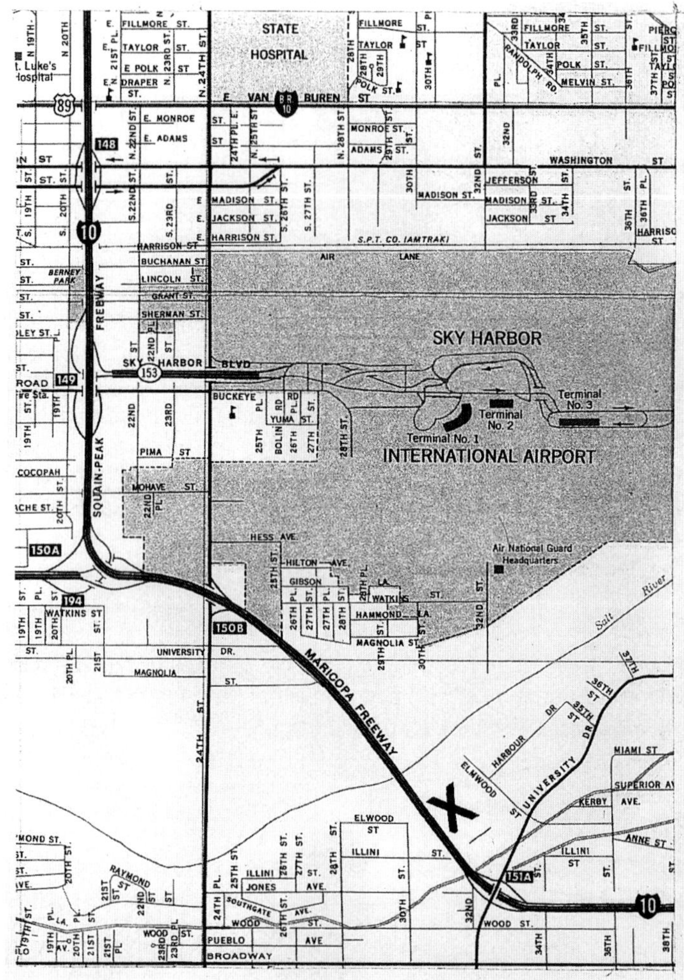

Fig. 2. Map indicating the location (bold X) of the urban riparian study area parallel to I-10 interstate highway in Phoenix, Arizona.

to a bimodal pattern with greater amounts falling during the winter months of December and January, and in the summer as thunderstorms during the monsoon period of July and August. Average annual precipitation is 194.6 mm (NOAA, 1996). The site selected for this study represents one of the few active riparian areas existing within the metroplitan area of Phoenix. Riparian and wetland vegetation spontaneously began colonizing the Tempe drain about 10 years ago, in a municipal stormwater channel located in the southeastern region of Phoenix. The drainage channel parallels the east side of interstate highway 10 (I-10) near the intersection of I-10 and University Drive (Fig. 2). The channel empties into the Salt River south of the Sky Harbor International Airport. The channel's banks are armored with large boulders, many greater than 1 m in diameter. The base of the channel contains smaller boulders with about a 10 cm deposition of fine sediment. Channel width is approximately 2 m, and the study area channel length was about 600 m. Permanent water in the channel collects from a variety of agricultural, residential, and industrial sources across the urbanized and mixed-use landscape that characterizes this portion of the Phoenix metropolitan area.

Vegetation sampling

Vegetation was sampled in late August and early September, 1993. The entire study area length was approximately 600 m, from which five 2 × 100 m length belt transects were established on each steambank. Sampling began at a point where a major drain under the I-10 freeway joined the main channel. At this point, woody plants became more common along channel until it joins the Salt River. Woody plant density was determined by direct count of individual species of the trees and shrubs along the channel. Canopy diameter was measured within an accuracy of 1 dm for each woody plant in the transects, and canopy cover determined by the product of canopy area and density for each species. Relative density was calculated to determine species composition. Plant height was estimated within 0.25 m of height using a calibrated pole. Canopy cover and density data were calculated for each transect and on a per hectare basis. Since no statistical differences were found between the transects on either bank, data were combined for presentation. While native to the Sonoran Desert in isolated relict communities, *Washingtonia filifera* (California fan palm) was included in the alien plant category when considering vegetation for this study, since it is not a component of riparian habitats of streams in the more natural parts of the region (Brock 1987, 1994a).

Results and discussion

Alien invasive woody plants

Morphological and ecological characteristics of nine of the ten alien invasive woody plants growing along the drainage channel will be discussed in the following section of this paper. *Tamarix chinensis* will not be included, as readers can find a description of this species in Brock (1994b). Information concerning these invasive species was compiled from Bailey 1949, Clark 1977, US Congress 1993, and Young and Young 1992.

Callistemon viminalis (Australian weeping bottlebrush) in the family *Myrtaceae* is a small tree or shrub with a height of about 6 to 10 m. This small tree has pendulous branches. The alternately arranged leaves are linear or lanceolate, narrow, light green rather stiff and about 15 cm in length (Fig 3a). Flowers are bright red clusters of stamens in dense spikes. Fruits are woody capsules that may persist on the stem for several years. This plant is drought tolerant but grows best in moist, well drained soils, and is also tolerant to saline-alkaline soils. This genus is closely related to *Melaleuca,* an invasive species of particular concern in Florida.

Ceratonia siliqua (carob) is a member of the *Fabaceae* family and was introduced to the United States in 1854 from the eastern Mediterrean. This small to moderate sized "evergreen" tree (3 to 15 m height) develops a dense crown of foliage. The alternately arranged leaves are even pinnately compound, with broad, shiny, leathery leaflets (Fig. 3b). In the Sonoran Desert, *C. siliqua* flowers in early winter, with the flowers in short clustered racemes borne on old wood. The flowers are a calyx tube with five short segments, have no petals, and five stamens. The fruit is an elongated coriaceous, leathery indehiscent pod (Fig. 3b). The pod is about 10 to 30 cm long and 0.6 to 2 cm thick. *C. siliqua* pods can be processed and used as a chocolate substitute.

Eucalyptus camaldulensis (red gum) is in the *Myrtaceae* family and can become a very large tree of up to 60 m. The tree was first introduced over 150 years ago to the United States in California from Australia. The flowers are in umbels or heads with five to ten flowers of many stamens. The petals are united into calyx tubes and the flowers are primarily insect pollinated. *E. camaldulensis* flowers in February through April with fruit ripening from July through October. The fruit is a capsule ending in an aburpt point or beak (Fig. 3c). The seeds are small with about 780 seeds/g. The thick, highly aromatic leaves are narrowly lanceolate in shape (Fig. 3c) and grayish green in color.

Leucaena glauca (lead tree) is a member of the *Fabaceae* family. *L. glauca* is native to tropical America, has no spines, and attains of mature height of about 10 m. The alternate leaves are bipinnately compound with eight or more paired pinnae of leaflets (Fig. 3d). The 3 cm diameter white flowers are in globous heads, usually bisexual with a five lobed calyx, five separated sepals, and ten exserted stamens. The fruits are sharp pointed pods 12 to 18 cm long and 1 to 2 cm wide. Pods are borne in clusters of fifteen to twenty with each pod containing a similar number of seeds. The pods easily dehisce the ellipitical, compressed shiny brown seeds. *L. glauca* seeds germinate well with seed coat scarification, and plants of two years of age have been reported to produce seed (Young and Young 1992). *L. glauca* is quite invasive within the urban landscapes of the Sonoran Desert. In the Sonoran Desert its use is as a shade tree, but in some parts of the world *L. glauca* is utilized as a fodder tree for livestock.

Melia azedarach (Texas umbrella, China berry, or bead tree) is a member of the *Meliaceae* family. This tree was introduced to the United States from southwest Asia. *M. azedarach* grows to about a 15 m height. Branches are smooth and reddish-brown and older branches and the trunk has furrowed bark. The 0.25 to 0.5 m sized leaves are bipinnately compound (Fig. 3e). The leaflets are 2 to 5 cm in length and have serrate or lobed margins. The flowers of *M. azedarach* have 5 to 6 sepals and petals, are white to violet in color, and appear in March through May. The fruit is a nearly globular drupe one to two cm in diameter, yellowish in color, and smooth. The fleshy mesocarp contains a stony center that contains four to five smooth but sharp pointed seeds. Large drooping branches and the foliage gives *M. azedarach* the appearance of a gigantic umbrella, which is impressive in urban landscapes.

→

Fig. 3a-i. Diagrams of the morphological characteristics of alien plants growing in an urban riparian habitat in the Sonoran Desert of south-central Arizona.

a ***Callistemon viminalis***

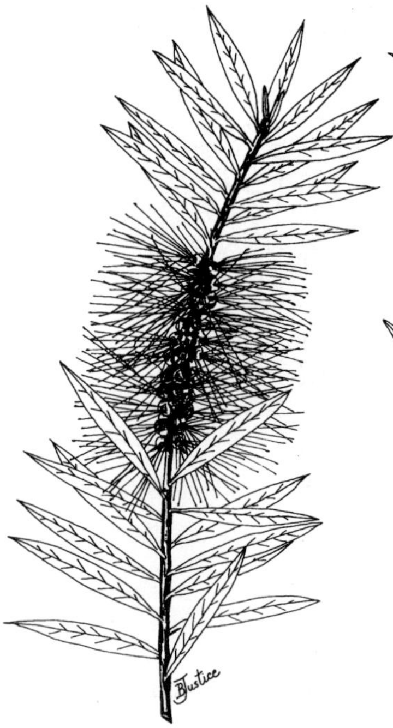

c ***Eucalyptus camaldulensis***

Ceratonia siliqua

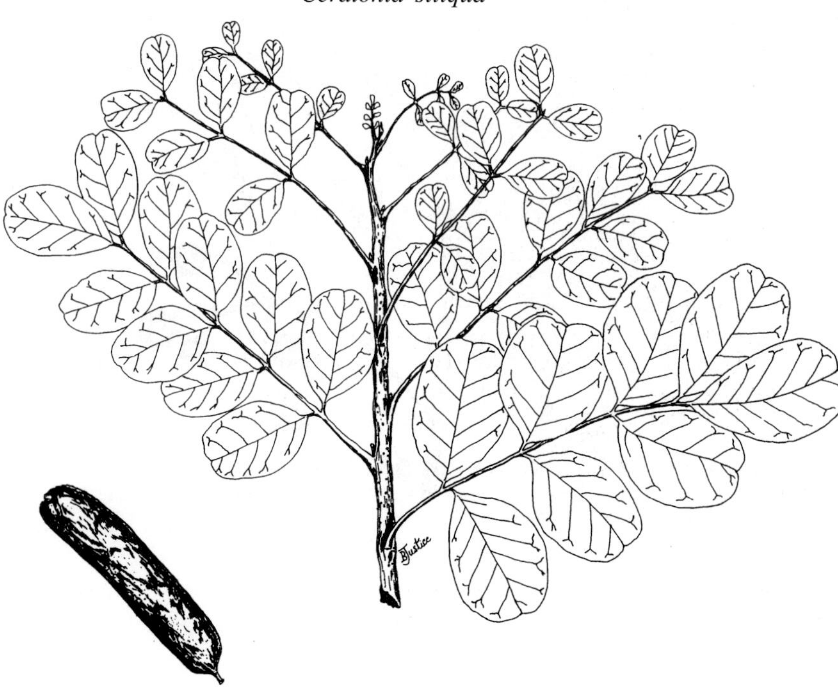

b

d *Leucaena glauca*

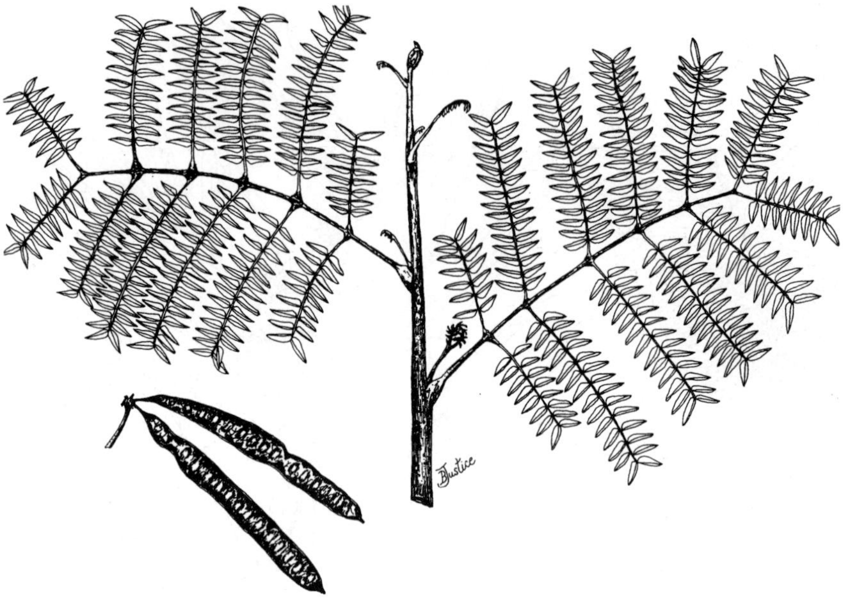

e *Melia azedarach*

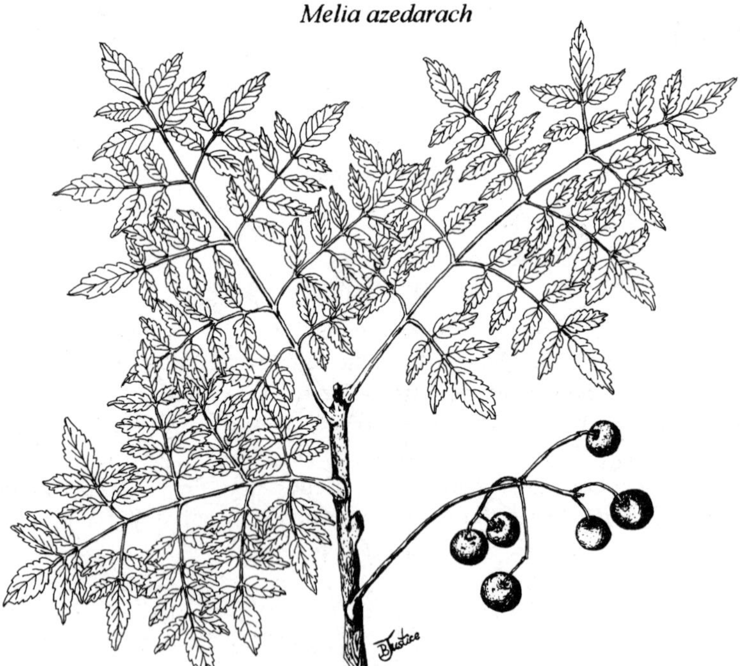

Fig. 3. d-e

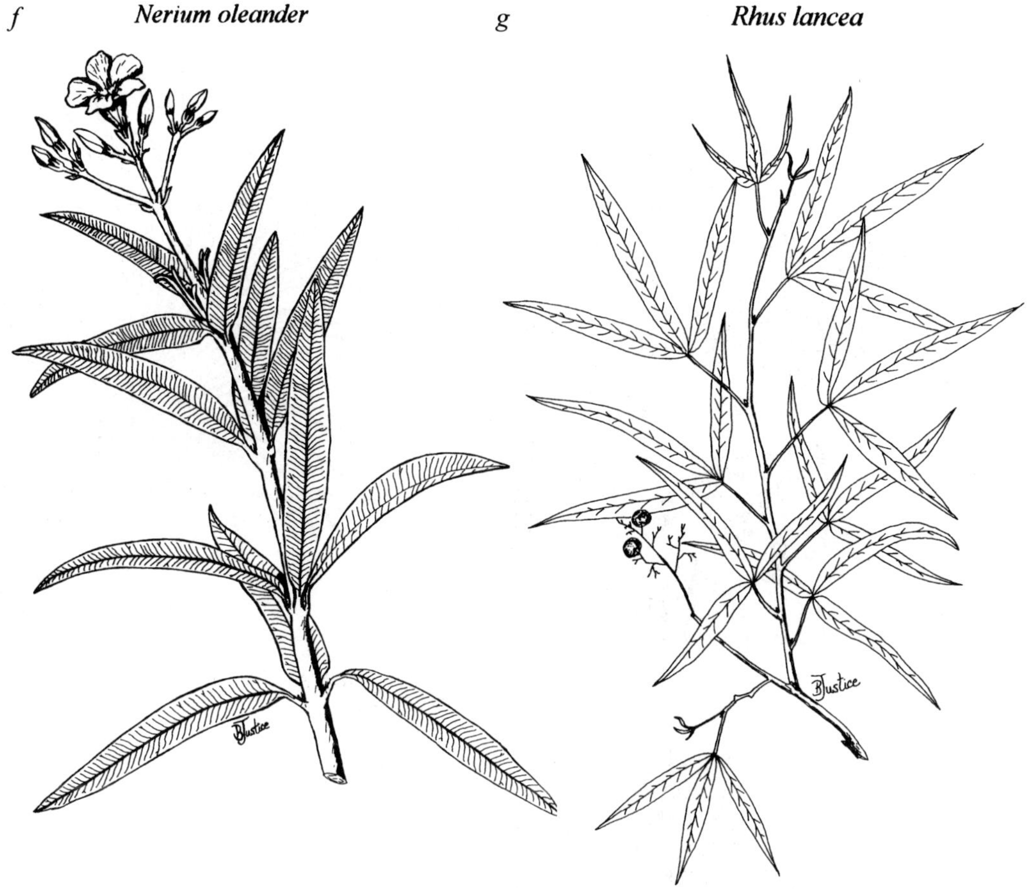

Fig. 3. f-g.

Nerium oleander (oleander) is an "evergreen" woody shrub in the *Apocynaceae* family that can attain a height of 6 m. The base of the plant is composed of many separate stems. This poisonous plant is native to the Mediterranean region. The dark green, lanceolate, leathery leaves are in whorls of three (Fig. 3f). The showy flowers are in terminal branching racemes. Typically the flowers are white or red in color, however, there are several horticultural varieties providing a range of colors. The calyx has glands at the base, the corolla is a funnel form, with the petals being 2 to 6 cm in width. The corolla is often double providing even more color to the landscape.

Rhus lancea (African sumac) is a slow growing tree of the *Anacardiaceae* family that can attain an 8 m height. The tree has interesting branching patterns with rough, dark reddish brown bark. The canopy is open with weeping outer branchlets. The compound leaves are trilobed, 8 to 12 cm long, with entire leaflet margins (Fig. 3g). The small green flowers are in axillary or terminal panicles and are produced in January. Small yellowish-red fruits are one seeded, with a sticky resinous coating borne in clusters. The plant "suckers" readily, and high recruitment of seedlings has been

h ***Ulmus parvifolia*** *i* ***Washingtonia filifera***

Fig. 3. h-i

observed by the author in flower beds under *R. lancea.* It is hypothesized that the seeds of *R. lancea* are dispersed by birds which would explain its presence in sites removed from other *R. lancea* trees. This tree species is now observed in many human influenced parts of urban landscapes.

Ulmus parvifolia (Chinese elm) of the *Ulmaceae* family grows rapidly often reaching a 6 m height within five years. The mature height of the tree is about 10 m. The tree is “evergreen” or deciduous depending on the severity of the winter in the Sonoran Desert. The younger branchlets are pubescent, but older branches are glabrous. As the trunk and limbs become older, the bark exfoliates leaving a smooth surface. The leaves are alternate, with unequal bases, thick and shiny, with serrate to crennate margins (Fig. 3h). The fairly inconspicuous flowers are bisexual in axillary clusters or racemes. The fruit is a disc shaped samara with a membraneous wing encircling the seed. Flowering and fruit development is in the late summer and autumn.

Washingtonia filifera (California fan palm) is a woody monocot in the *Arecaceae* family. The tree can reach a height of 15 m, capped by a display of green material

Table 1. A list of woody plant species growing in a Sonoran Desert urban riparian habitat, Phoenix, AZ.

Scientific name	Family	Common name
Alien species:		
*Nerium oleander**	*Apocynaceae*	oleander
Callistemon viminalis	*Myrtaceae*	weeping bottle brush
Ceratonia siliqua	*Fabaceae*	carob
Eucalyptus camaldulensis	*Myrtaceae*	red gum tree
Leucaena glauca	*Fabaceae*	lead tree
Melia azedarach	*Meliaceae*	Texas umbrella tree
Rhus lancea	*Anacardiaceae*	African sumac
Tamarix chinensis	*Tamaricaceae*	salt cedar
Ulmus parvifolia	*Ulmaceae*	Chinese elm
Washingtonia filifera	*Arecaceae*	California fan palm
Native species:		
Cercidium microphyllum	*Fabaceae*	foothills paloverde
Fraxinus velutina	*Oleaceae*	velvet ash
Populus fremontii	*Salicaceae*	Fremont cottonwood
Prosopis velutina	*Fabaceae*	velvet mesquite
Salix gooddingii	*Salicaceae*	Goodding willow
*Baccharis sarathoides**	*Asteraceae*	desert broom
*Baccharis salicifolia**	*Asteraceae*	seep willow

* Shrub species

underlaid by groups of dry leaves that forms a thatch like shroud that may remain on the trunk for many years. The trunk tapers slowly from the base to the top, and when the dry material is shed the trunk becomes smooth. The crown of the tree is made up of perhaps 50 fan shaped leaves (fronds) with petioles as long as 1.5 m in length (Fig. 3i). The petiole margin is lined with saw-like, sharp, bony projections. The fronds are fan shaped and about 1 m in width. The leaves are folded, cut nearly to the middle, and often bear conspicuous sting-like filaments. In June *W. filifera* produces long thin spathes, from which creamy colored flowers emerge in panicle-like spadices. The fruit matures in late summer/autumn and is a small ellipsoid black drupe. A fleshy layer surrounds a single seed, and the fleshy material serves as food for birds. *W. filifera* is not an alien plant to the hot southwestern deserts of North America. It is native to southern California, northern Baja California in Mexico, and in Yuma county Arizona. In these areas is exists naturally as relict vegetation in isolated canyons, and is not a component of modern Sonoran Desert riparian plant communities.

Vegetation structure

Ten of the seventeen woody plants (59 %) documented in the vegetation sampling of the urban channel are considered alien to the normal riparian flora of the Sonoran Desert. Those woody species are listed in Table 1. One alien shrub (*N. oleander)* was encountered in the sampling with the other nine alien plants have a tree life form. The most common alien trees were *W. filfera* and *T. chinensis*, with *Cercidium microphyllum* (palo verde) and *Salix gooddingii* (Goodding willow) being the most com-

Table 2. Average tree density (number/ha), percent relative density, percent canopy cover, and plant height (m) of native and alien trees in a Sonoran Desert urban riparian habitat in Phoenix, AZ.

Vegetation character	Native trees	Alien trees
Density (number/ha)	175	425
Relative density (%)	26.9	73.1
Canopy cover (%)	24.1	38.6
Height (m)	4.9	3.9

mon native trees. Of the alien trees, *C. viminalis, E. camaldulensis,* and *R. lancea* were intermediate in presence while *C. siliqua, L. glauca, M. azedarach,* and *U. parvifolia* were fairly rare. Only one *Populus fremontii* (Fremont cottonwood) was found on the site, with *Fraxinus velutina* (velvet ash) and *Prosopis velutina* (velvet mesquite) being native trees with intermediate presence.

Height of all trees averaged 4.4 m. The alien shrub, *N. oleander* averaged 1.5 m in height. Native trees averaged 1 m greater height compared to average alien tree height (Table 2). The average height of the native trees was 4.9 m (Table 2), with the greatest tree height being measured in the center part of the study area (Fig. 4). The average height of alien woody plants was 3.9 m, with the greatest average height, 4.8 m also being in the center section of the site (Fig. 4). Only in the segment of the channel closest to the Salt River (transect 5), did alien woody species have height greater than the native species. While native species were fewer in number, they had better height development than the alien species. The main height contibution was from *P. fremontii, S. gooddingii,* and *C. microphyllum.*

There were more alien trees along the stream than native trees except for the transect nearest the Salt River (Fig. 5). The greatest density of alien and native plants (50 trees) occurred in the second sample area where it has been hypothesized that nutrient uptake by the riparian species was greatest (Farkas 1994). Another explanation for the greater density of trees in this section of the channel could be from seed loading. This section of the channel begins at a 100 m distance from a discharge source to the channel and the greater number of plants could be from seed deposition. This discharge point is a large drain pipe that conveys water from the south and west sides of the I-10, known as Maricopa highway (Fig. 2). Woody plant density decreased rapidly downstream with only three alien plants being found in the last section of the channel before it joins the Salt River (Fig. 5).

Of all the alien plant species invading the channel banks, *W. filifera,* classified for this study as an alien tree, had the greatest relative density, about 42 % (Fig. 6). *T. chinensis* had just more than a 20 % relative density value, with *C. serotina, L. glauca* and *M. azedarach* being found only once on the site (Table 3, Fig. 6). The most common alien tree was *W. filifera*, while *S. gooddingii* was the most common native tree species.

Canopy cover along the channel was greatest in the second transect area. Total canopy cover of 170 % was calculated for that channel segment (Fig. 7). There was considerable canopy overlap in this portion of the channel, but only about 20 % of the total canopy cover was contributed by native tree species. In rural riparian habitats of central Arizona, canopy cover in mature stands exceeded 100% at several sites, as was measured in this urban riparian habitat, however the only alien tree

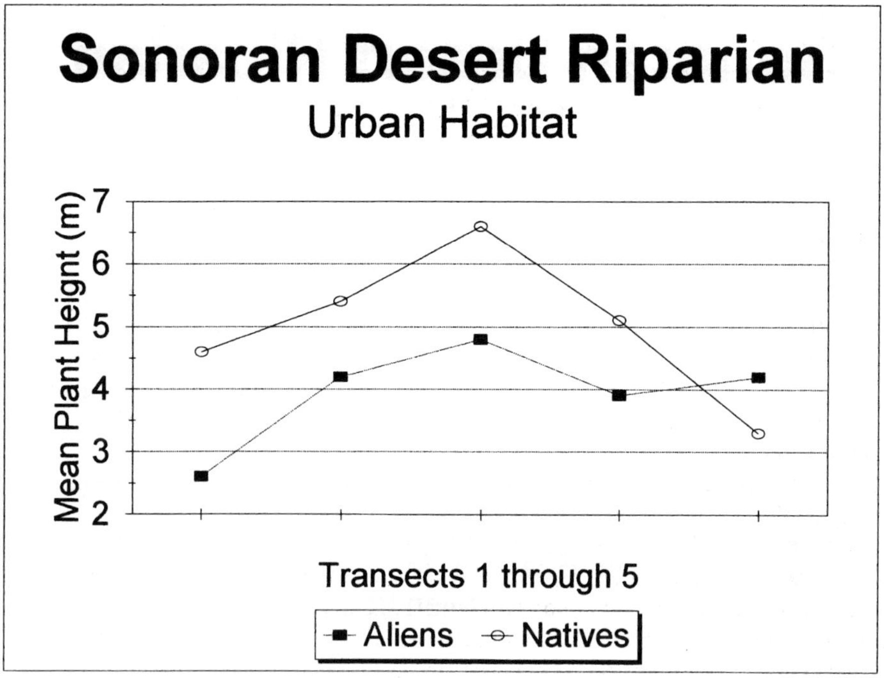

Fig. 4. Average plant height (m) for alien and native woody tree species in a Sonoran Desert urban riparian habitat, Phoenix, Arizona, USA.

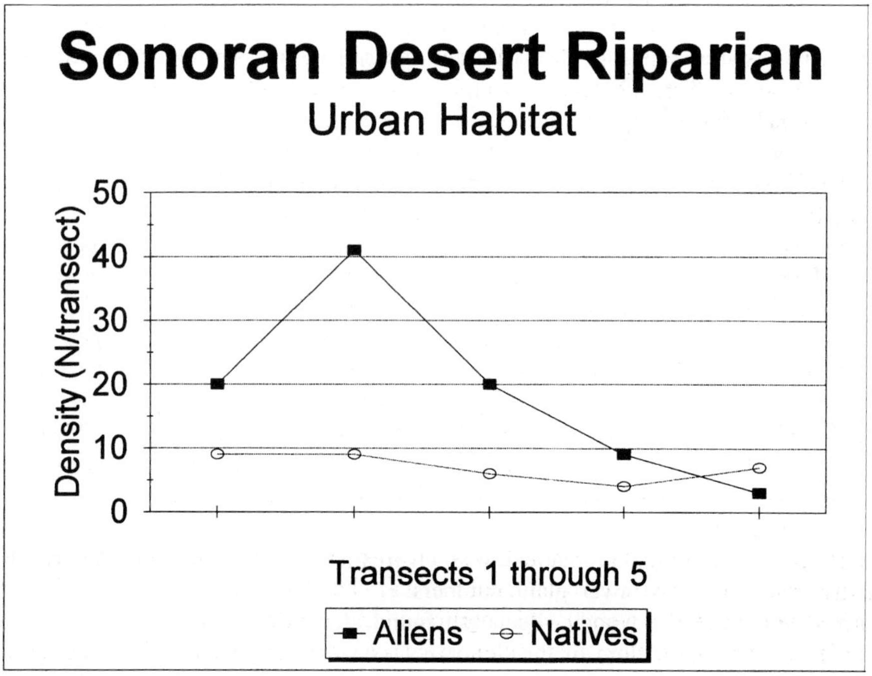

Fig 5. Average density (N/200 m^2) for alien and native woody tree species in a Sonoran Desert urban riparian habitat, Phoenix, Arizona, USA.

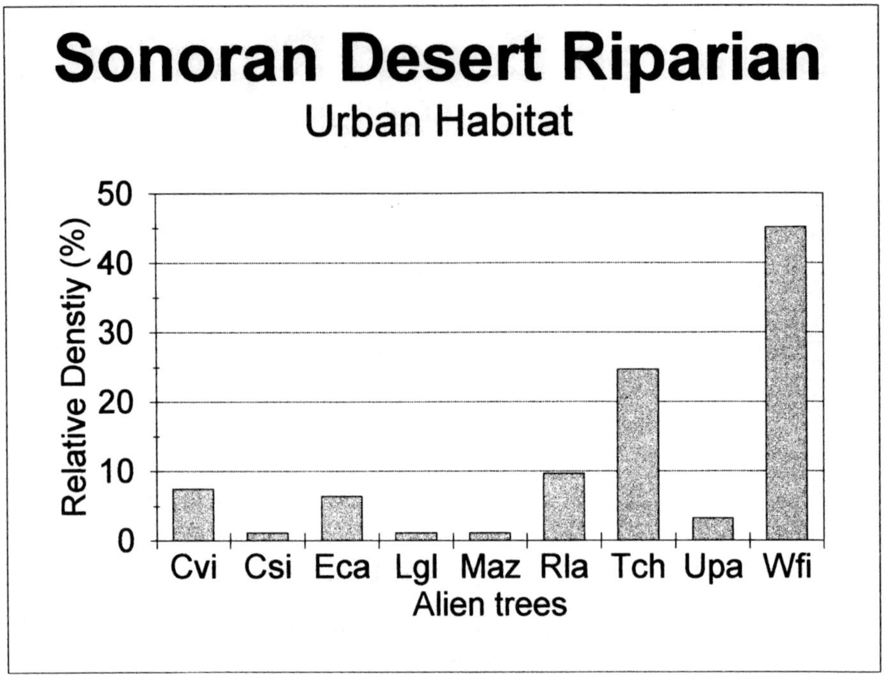

Fig. 6. Relative density (%) of alien riparian trees growing in an urban Sonoran Desert habitat, Phoenix, Arizona, USA.

Table 3. Relative density (%) for alien and native tree species in a Sonoran Desert urban riparian habitat, Phoenix, Arizona, USA.

Tree Species	Relative Density (%)	Tree Species	Relative Density (%)
Alien trees:			
C. viminalis	5.3	*C. siliqua*	0.7
E. camaldulensis	4.6	*L. glauca*	0.7
M. azedarach	1.4	*R. lancea*	6.9
T. chinensis	17.8	*U. parvifolia*	3.1
W. filifera	32.3	Alien trees total	72.8
Native trees:			
C. microphyllum	8.5	*F. velutina*	3.1
P. fremontii	0.7	*P. velutina*	4.6
S. gooddingii	10.0	Native trees total	26.9

encountered in those studies was *T. chinensis* (Brock 1987; Brock 1994a). Tree canopy cover decreased dramatically downstream and was not only the result of smaller trees, but also fewer plant numbers.

Sixty percent of the woody vegetation found along this urban riparian study area are alien to the native flora of the Sonoran Desert. In another riparian area (ephemeral stream flow) of the Sonoran Desert that had a gradient from urbanized to wildlands, Green and Baker (1997) found that 27 % of the plants were alien, with

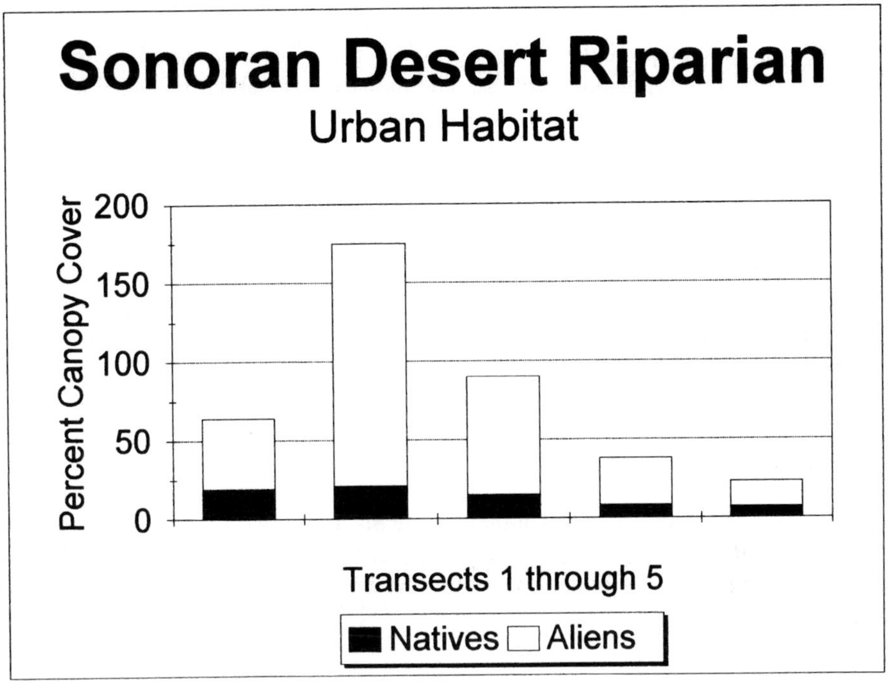

Fig. 7. Average canopy cover (%) for alien and woody trees in a Sonoran Desert urban riparian habitat, Phoenix, Arizona, USA.

R. lancea and *W. robusta* (Mexican fan palm) representing the woody life-form. In Stromberg *et al.* (1997) it is reported that 23 % of the herbaceous species found in the Sonoran Desert riparian areas studied were exotics. While alien (exotic) plants are commonly found along streams in the Sonoran Desert the degree to which they may become invasive largely remains to be determined.

Conclusion

The alien woody plants observed along this urban habitat in the Sonoran Desert are facultative riparian species rather than obligates. These species persist in this arid climate with little additional water, although their presence in landscapes where they would not benefit from some run-in water is questionable. However, in areas where water naturally collects these species, once established, would probably maintain themselves. These alien riparian plants have seeds and seed dispersal mechanisms that favors transport with flood debris, or zoochory. Additionally, the hard seeded species, *L. glauca, C. siliqua, W. filifera,* and *R. lancea* have seeds that may benefit from mechanical scarification when transported with the sand and gravel bedload common to the streams of arid climates, and result in higher germination success in riparian areas compared to upland sites. The seeds of the remaining species could easily reach the riparian area through wind transport.

Early warning of the invasive nature of plants can use the process of converging evidence, and information presented in this paper provides evidence for several species that have not previously been documented as invaders to riparian habitats in the southwestern United States, although some have been observed by the senior author in upland anthropogenic habitats. The moist linear habitat provides less resistance for conveyance of propagules, and the exotic species' emerging invasive characteristics should cause concern among landscape ecologists, and persons managing stream corridors.

Of the ten alien woody species growing in the stream channel, the invasive nature of *T. chinensis* is well known. Based on observations of the senior author, the alien plants growing in the urban riparian habitat having the highest degree of invasive potential are *L. glauca* and *R. lancea*. Species with moderate invasive potential are *U. parvifolia, E. camaldulensis,* and *N. oleander,* while those with less invasive potential include *M. azedarach, C. siliqua*, and *C. viminalis*.

Acknowledgments

We would like to thank Chandra Chithalura for his assistance in vegetation sampling. Drawings of the alien plant morphological characteristics are the work of Brenda Justice. This research was funded in part by the Southwest Center for Environmental Research and Policy.

References

Bailey, L. H. 1949. Manual of Cultivated Plants. MacMillan Co., New York, NY.

Bazzaz, F. A. 1986. Life history of colonizing plants: Some demographic, genetic, and physiological features. In: H. A. Mooney and J. A. Drake, (eds.), Ecology of Biological Invasions of North America and Hawaii. Ecological Studies 58, pp. 96-110. Springer-Verlag, New York.

Brock, J. H. 1987. Structure of riparian habitats at selected sites along the Verde and East Verde rivers of central Arizona. Report to US Bureau of Reclamation. Arizona State University, Tempe, AZ

Brock. J. H. 1994a. Phenology and stand composition of woody riparian plants in the southwestern United States. Desert Plants 11:23-32.

Brock, J. H. 1994b. *Tamarix* spp. (Salt Cedar), an invasive exotic woody plant in arid and semi-arid riparian habitats of western USA. In: L. C. de Waal, L. E. Child, P. M. Wade, and J. H. Brock, (eds.), Ecology and Management of Invasive Riverside Plants. J. Wiley and Sons, Chichester, UK.

Clark, D. E. (ed). 1977. Western Garden Book. Lane Publishing Co., Menlo Park, CA.

Farkas, M. C. 1994. Wetland and riparian function and effectiveness in the mitigation of urban nonpoint source pollution. MS Thesis. Arizona State University.

Green, D. M. and M. G. Baker. 1997. Species composition along a gradient of urbanization in the lower Sonoran Desert, Arizona, USA. This volume.

Kowarik, I. 1995. Time-lags in biological invasions with regard to success and failure of alien species. In: P. Pysek, K. Prach, M. Rejmanek, and M. Wade, (eds.), Plant Invasions: General Aspects and Special Problems, pp. 15-38. SPB Academic Publ., Amsterdam.

National Oceanic and Atmospheric Administration. 1995. Climatological statistics for Arizona. US Department of Commerce.

Natural Resources Conservation Service. 1995. Major land resource areas and resource units, Arizona. USDA, NRCS, National Cartography & GIS Center, Fort Worth, TX.

Pimentel, D. 1986. Biological invasions of plants and animals in agriculture and forestry. In: H.A. Mooney and J. A. Drake, (eds.), Ecology of Biological Invasions of North America and Hawaii. Ecological Studies 58, pp. 149-162. Springer-Verlag, New York.

Pyšek, P., K. Prach, M. Rejmanek, and M. Wade. 1995. Plant Invasions: General Aspects and Special Problems. SPB Academic Publ., Amsterdam.

Pyšek, P., K. Prach, and P. Smilauer. 1995. Relating invasion success to plant traits: An analysis of Czech alien flora. In: P. Pysek, K. Prach, M. Rejmanek, and M. Wade, (eds.), Plant Invasions: General Aspects and Special Problems, pp. 39-60. SPB Academic Publ., Amstersdam.

Rejmanek, M. 1995. What makes a species invasive? In: P. Pyšek, K. Prach, M. Rejmanek, and M. Wade, (eds.), Plant Invasions: General Aspects and Special Problems, pp. 3-13. SPB Academic Publ., Amsterdam.

Roughgarden, J. 1986. Predicting invasions and rate of spread. In: H. A. Mooney and J. A. Drake, (eds.), Ecology of Biological Invasions of North America and Hawaii. Ecological Studies 58, pp.179-188. Springer-Verlag, New York.

Stromberg, J. C., L. Gengarelly and B. F. Rogers. 1997. Exotic herbaceous species in Arizona's riparian ecosystems. This volume.

US Congress. 1993. Harmful Non-indigenous Species in the United States. Office of Technology Assessment. OTA-F-565. US Government Printing Office. Washington, DC.

Vitousek, P. M. 1986. Biological invasions and ecosystem properties: Can species make a difference? In: H. A. Mooney and J. A. Drake, (eds.), Ecology of Biological Invasions of North America and Hawaii. Ecological Studies 58, pp.163-176. Springer-Verlag, New York.

Vitousek, P. M., C. M. D'Antonio, L. L. Loope and R. Westbrooks. 1996. Biological invasions as global environmental change. American Scientist 84:468-478.

Young, J. A. and C. G. Young. 1992. Seeds of Woody Plants in North America. Dioscorides Press. Portland, OR.

SPECIES COMPOSITION ALONG A GRADIENT OF URBANIZATION IN THE LOWER SONORAN DESERT, ARIZONA, USA

Douglas M. Green* and Michael G. Baker[1]
[1] *School of Planning and Landscape Architecture, Arizona State University, Tempe AZ 85287-2005, USA*; * *corresponding author*

Abstract

Urban expansion into wildlands has reduced and fragmented formerly contiguous habitats. Embedded within the urban matrix are newly formed habitat islands in a fragmented riparian ecosystem. Although these islands are riparian in floristic and physiognomic character, their function as habitat is not well understood. We assessed plant species composition of riparian habitats along a gradient of increasing urbanization within the city limits of Phoenix, Arizona, USA. Herbaceous plant composition was quantified by species frequency and cover. Woody vegetation was quantified on the basis of density, height, and canopy volume. We identified a total of 69 species of which 19 or 27% were introduced. Urban communities had significantly higher richness of introduced shrubs and trees. Overall diversity (H') was 2.0674. The urban wash site had the highest total species richness and diversity, however these values were inflated by a larger component of introduced species. Diversity ranged from 2.1680 in the urban sites to 1.6725 in the margin sites and 2.0566 in the rural sites. Volume and density of tree and shrub canopies were significantly higher in the margin communities. We feel the impacts of urbanization upon riparian habitats could be reduced by proper urban planning and implementation of active habitat management strategies aimed at maintaining landscape mosaics and function.

Introduction

Rural landscapes have changed dramatically in recent decades as a result of residential development emanating from rapidly growing urban centers. Residential development acts to fragment once contiguous regional landscapes. Habitat fragmentation occurs when a large, continuous vegetation type is converted to other uses or vegetation types such that only scattered fragments of original type remain. These habitat remnants (also referred as isolates, habitat islands, fragments, patches) occupy less area, are of variable size, shape and location, and separated from the original habitat (MacArthur and Wilson 1967; Brown and Gibson 1983; Faaborg *et al.* 1992). Habitat fragmentation has been associated with quantitative and qualitative loss of habitat for species originally dependent on that habitat type thus leading to declines in diversity of species present within the fragments (Temple and Wilcox 1986).

Fragmented riparian environments of similar size and vegetative structure may not be ecologically equal because of differences in their surroundings (Freisen *et al.* 1995). Remnant habitat fragments or patches are subject to many adverse effects, relating to matrix-patch relationships (Forman and Godron 1986; Saunders *et al.*

Plant Invasions: Studies from North America and Europe, pp. 37–44
edited by J.H. Brock, M. Wade, P. Pyšek and D. Green

1991). Insular effects within habitats frequently lead to changes in native disturbance regimes, physical environment, and loss of native vegetation through encroachment by exotic species (Temple and Wilcox 1986; Stromberg *et al.* 1992). Though development of high density housing is not inherent to the loss in riparian vegetation and abiotic habitat in small patches, their deterioration is driven by external rather than internal forces (Saunders *et al.* 1991; Engels and Sexton 1994). In addition, faunal communities within urban habitats are adversely effected by the lack of coevolved interspecific relationships, contributing to the rapid demise of native floral and faunal communities (Ambuel and Temple 1983; Johnson and Temple 1986; de Waal *et al.* 1994). The objective of this study was to quantify plant species richness and diversity along a gradient of urbanization in the upper sonaran desert.

Study area

Our study was conducted within the city of Phoenix located in south central Arizona. The City of Phoenix is a rapidly expanding urban center, most expansion is in the form of single family housing. Study areas were located along Cave Creek Wash in the upper sonaran desert vegetation type. Cave Creek is an approximately 75 km long ephermal wash. Cave Creek drains natural habitats north of phoenix and becomes increasingly modified until the last 10 km where subsurface storm drains carry flows to the Salt River, the major regional river. Elevation of the study reach is about 382 m. Average winter temperature is 11.3°C and average summer temperature is 30.8°C. Average annual precipitation is 200 mm of which about 50% occurs as summer thunderstorms (Sellers and Hill 1974). Study areas were selected to represent a gradient of urbanization and range in size from 14,643 m^2 to 23,550 m^2 (Table 1). We divided each study area into three sites for vegetation description.

Methods

We sampled woody perennial vegetation with six 2×20 m belt transects for shrubs and 50 m radius plots to sample trees. Species considered shrubs were measured within 2 m^2 subplots, for crown density, maximum height, and radius by species of shrubs originating within the subplot (Brown 1982; Bonham 1989). Tree species sampled within three 50 m radius plot, were quantified for density, measured for basal circumference, maximum height and radius of the canopy. We sampled herbaceous species at each site with six 30 m transects. A 20×50 cm plot was placed at one meter intervals, and canopy coverage of species was estimated using cover classes (Daubenmire 1968). All vegetation was sampled during the spring of 1995. We used aerial photographs (1:1200) to estimate housing density and aerial dimensions of each study site (Landiscor 1995).

Species diversity was calculated using the AID program (Overton et al 1987). We used the Shannon-Weaver index (H') as our measure of diversity (Shannon and Weaver 1949). Species richness and diversity between sites were compared as dependent variables in the formal analysis. The population variables were compared using Kruskal-Wallis procedure (Zar 1984). Significance of the various parameters were tested at minimum $p = 0.05$.

Table 1. Mean house density (# ha^{-1}) and study area size (m^2) for three habitats along the Cave Creek Wash.

Habitat	House Density	Study area
urban	7.2±1.4[1]	14,923
margin	2.1±0.1	14,643
rural	0	23,550

[1]Standard error of the mean

Results

Urban wash sites

Urban wash sites had a total of 56 species of which 37% were non-native. The most common herbaceous species was the introduced species *Bromus rubens* (red brome) with a relative frequency of 51% (Table 2). The most common native species was *Xanthium saccharatum* (common cocklebur) with a relative frequency of 14%. Species richness of shrubs was 19 and species diversity was 1.1483. Average stem density of shrubs was 48 ha^{-1} and average shrub height was 1.7 m (Table 3). *Baccharis sarathroides* (desert broom), a native species had the highest stem density (98 ha^{-1}) and a average height of 2.41 m. Urban wash sites had siginficantly larger shrub canopy volume than either margin or rural sites (Table 3). We found 2 introduced shrub species *Cassia* spp. (Cassia) and *Dasylirion wheeleri* (desert spoon), both escapees from cultivation. Ten tree species were identified in the urban sites. Average tree density was 7.9 ha^{-1} with an average height of 5.1 m (Table 4). Introduced trees include escapees from cultivation *Rhus lancea* (African sumac) and *Washingtonia robusta* (Mexican fan palm); these were encountered infrequently. *Tamarix chinensis* (salt cedar) was found only in the urban site and had a density of 7.9 ha^{-1}. The native species *Cercidium floridum* (blue palo verde) was the most commonly occurring species with a density of 4.5 ha^{-1}.

Margin wash sites

Margin wash sites had a total species richness of 38 and a species diversity of 1.6725 (Table 2). These sites were dominated by shrubs with a relatively small herbaceous component. Twenty-one percent of the herbaceous species found were not native to the Sonoran desert region. The most common introduced species was *B. rubens* with a relative frequency of 50%. The most commonly occurring native species was *Sphaeralcea ambigua* (desert mallow) with a relative frequency of 11%. Shrub density in the margin sites was 421 ha^{-1}, significantly higher than either the margin or rural sites (Table 3). Average shrub density was 1.8 m. Shrub volume was significantly higher than either urban or rural sites (Table 3). No introduced shrub species were found in the margin sites. *Ambrosia deltoidea* (triangle leaf bursage) and *Larrea tridentata* (creosote bush) has the highest stem densities, 129 and 123 ha^{-1} respectively. Average tree density in the margin sites was 27 ha^{-1}, significantly higher than either the urban or rural wash sites (Table 4). The most commonly occurring tree was *Cercidium floridum* with an average height of 5.4 m. No exotic trees were encountered during sampling.

Table 2. Native and introduced species richness and total species diversity (H') for urban, margin, and rural sites along Cave Creek Wash.

	Urban			Margin			Rural		
	richness			richness			richness		
Life form	total	exotic	diversity	total	exotic	diversity	total	exotic	diversity
tree	5	4	1.0986	4	0	0.9437	5	0	1.1793
shrub	12	2	1.1483	11	0	1.5656	15	0	1.6704
herb	22	11	1.9430	15	8	1.5467	20	12	1.9831

Table 3. Density (# ha^{-1}) and volume (m^3 ha^{-1}) of shrub species for urban, margin, and rural sites along Cave Creek Wash. Values are means ± standard deivation.

	Density		Volume	
Habitat	Native	Introduced	Native	Introduced
urban	48±14.9^a	14±8.1	14,011±2,151^a	428.2±93.0
margin	421±96.2^b	0	11,737±878^b	0
rural	272±71.6^a	0	5,472±3,424^a	0

Entries followed by different letters in the same column are significantly different at p=0.05.

Table 4. Density (# ha^{-1}) and volume (m^3 ha^{-1}) of tree species for urban, margin, and rural sites along Cave Creek Wash. Values are means ± standard deivation.

	Density		Volume	
Habitat	Native	Introduced	Native	Introduced
urban	7.9±4.9^a	2.3±2.2	580±305^a	107.2±53.0
margin	27±7.5^b	0	8,456±1416^b	0
rural	13±2.5^a	0	2,659±273^a	0

Entries followed by different letters in the same column are significantly different at p=0.05.

Rural wash sites

Rural wash sites had a total species richness of 52 and a species diversity of 2.0466 (Table 3). Twenty-three percent of herbaceous encountered were native to the region. The most common introduced species was *B. rubens* with a relative frequency of 30%. *Hordeum arizonicum* (wild barley) was the most common native species with a relative frequency of 15%. Average shrub density of rural wash sites was 272 ha^{-1} as in the margin sites no introduced shrub species were found (Table 3). Shrub stem density was highest for *A. deltoidea* and *L. tridentata* (452 and 100 stems ha^{-1} respectively). Average tree density in rural wash sites was 13 ha^{-1} with an average height of 4.7 m (Table 4). *C. floridum* was the most common tree species with a density of 14.7 ha^{-1} and an average height of 4.6 m.

Discussion

During our study we identified a total of 69 species of which 19 or 27% were introduced. Overall species diversity was 2.0674 (Appendix 1). Urban wash sites had the highest total species richness and diversity followed by rural and margin sites (Table 2). Species richness and diversity in urban wash sites reflected increased numbers of introduced species at this site. Several of these species were ornamentals. In all three sites introduced herbaceous species made up a significant component of the species richness. Many of these species such as *B. rubens* are ruderal species (sensu Grime 1988) that have life histories adapted to conditions of frequent disturbance. Urban sites have greater disturbance associated with human use than either margin or rural sites. This disturbance includes extensive chanelization and lining (rip-rap) of the wash channel. Other disturbance factors include flow alteration caused by channel and watershed modification, other human activies such as tree cutting and recreational use also influence species composition. In addition to physical disturbances, urban wash sites are also impacted by introduction of seeds and plant materials from adjacent communities either by natural vectors such as wind or animals or by dumping of yard trimmings directly into the wash.

Lining of wash channels also provides ideal habitat for the native shrub *B. sarathroides*. The upper bank areas of lined washes supported an almost monospecific stand of *B. sarathroides*. Lower channel banks were either devoid of vegetation of were dominated by annual species as infrequent high flows remove most perennial species. Encroachment by exotic shrub and tree species appears directly related to urban encroachment by their absence in margin and rural sites (Table 2). However the degree of encroachment in terms of density and volume of exotics appears to be minimal at current levels urbanization. This may be due to the long lived nature of Sonoran desert shrub and tree species relative newness of the urbanized areas (generally less than 20 years old), and infrequent opportunities for establishment of riparian shrubs and trees.

Urban sprawl is widespread throughout southwestern United States and threatening ecosystems through dewatering of aquifers, altering drainage patterns, and removal of native vegetation from the regional landscapes. As development reaches into rural areas, additional desert landscapes will be fragmented. Habitat loss and species shifts because of negative influences associated with nearby urban development is expected.

Additional research at the same study site is showing that effects of development severely reduce the ecological value of adjacent riparian habitats to bird communities. We stress the importance of preserving landscape mosaics as keeping the riparian vegetation intact, is itself inadequate to maintaining the native riparian function (Forman 1995).

Development and adoption of a comprehensive plan, incorporating wildlife values, the regulation of development and other land uses, acquisition of conservation easements or purchase of high-value parcels, and accommodation of the needs of nearby urbanites has recently become a concern to some Maricopa County residents (Ingley and Golfen 1995). Recent attitude surveys, conducted by Arizona State Parks, report that 94% of residents support not only recreational opportunities but the preservation of undeveloped natural areas. This suggests that certain urban riparian areas may deserve restoration and protection under state jurisdiction using state funds within municipal boundaries (Arizona State Parks 1992). To meet public demands for additional open spaces as undeveloped natural areas, management efforts should begin with a survey of its natural resources, in this case of native avian communities, to provide multidisciplinary planning teams baseline data for consideration into comprehensive planning for protection of remnant natural areas (Shaw *et al.* 1986).

Acknowledgements

This study was made possible by a grant from the Arizona Game and Fish Department Heritage Program, grant number U94016. We also acknowledge assistance from Maricopa Flood Control District and the City of Phoenix. The authors would like to thank the efforts of numerous volunteers who assisted in data collection. We specifically acknowledge the assistance of Jeanine Baker in many phases of this study.

References

Ambuel, B. and S.A. Temple. 1983. Area-dependent changes in the bird communities and vegetation of southern Wisconsin forests. Ecology 64:1057-1068.

Arizona State Parks. 1992. The 1992 Arizona outdoor recreation needs survey: Final report. Arizona State Parks, Phoenix, AZ.

Bonham, C.D. 1989. Measurements for Terrestrial Vegetation. John Wiley & Sons, New York, NY.

Brown, D.E. 1982. Biotic communities of the American Southwest-United States and Mexico. Desert Plants 4:1-342.

Brown, J.H., and A.C. Gibson. 1983. Biogeography. C.V. Mosby Company, St. Louis, MI.

Daubenmire, R. 1968. Plant communities: A textbook of plant ecology. Harper and Row, New York.

de Waal, L.C., Child, L.E., Wade, P.M. and Brock, J.H. 1994. Ecology and Management of Invasive Riverside Plants. John Wiley & Sons, New York, NY.

Engels, T.M. and C.W. Sexton. 1994. Negative correlation of Blue Jays and Golden-cheeked Warblers near an urbanizing area. Cons. Biol. 8:286-290.

Faaborg, J., M. Brittington, T. Donovan and J. Blake. 1992. Habitat Fragmentation in the Temperate Zone: A Perspective for Managers. In: Status and Management of Neotropical Migratory Birds. pp. 331-339. U.S.D.A., Forest Service, Rocky Mountain Forest and Range Experiment Station, Fort Collins, CO. Gen. Tech. Rep. RM-229.

Forman, R.T.T. 1995. Land Mosaics: The ecology of landscapes and regions. Cambridge University Press, New York.

Forman, R.T.T. and M. Godron. 1986. Landscape ecology. John Wiley & Sons, New York, NY.

Friesen, L.E., P.F.J. Eagles, and R.J. Mackay. 1995 . Effects of residential development on forest-dwelling Neotropical migrant songbirds. Cons. Biol. 9:1408-1414.

Grime, J.P., J.G. Hodgson, and R. Hunt. 1988. Comparative plant ecology: A functional approach to common British species. Allen and Unwin, Boston.

Ingeley, K. and B. Golfen. 1995. Can growing valley still preserve desert? pp.1 & 13-14 in The Arizona Republic, April 16, 1995, Phoenix, AZ.

Johnson, R.G. and S.A. Temple. 1986. Assessing habitat quality for birds nesting in fragmented tallgrass prairie. In: J. Verner, M.L. Morrison, and C.J. Ralph (eds.), Wildlife 2000: Modeling Habitat Relationships of Terrestrial Vertebrates. pp. 245-250. University of Wisconsin Press, Madison, WI.

Landiscor Aerial Surveys, Inc. 1969-1995. 810 N. 2nd St., Phoenix, AZ 85004.

MacArthur, R.H. and E.O. Wilson. 1967. The theory of island biogeography. Princeton University Press, Princeton, NJ.

Overton, W.S., B.G. Smith and C.D. McIntire. 1987. AID Programs: Analysis of Information and Diversity. Oregon State University, Corvallis, OR.

Saunders, D.A., R.J. Hobbs, and C.R. Margules. 1991. Biological consequences of ecosystem fragmentation: a review. Cons. Biol. 5:18-32.

Sellers, W.D., and R.H. Hill. 1974. Arizona climate 1931-1972. University of Arizona Press, Tucson.
Shannon, C.E., and W. Weaver. 1949. The mathematical theory of communication. University of Illinois Press, Urbana, IL.
Shaw, W.W., J.M.Burns, and K. Stenberg. 1986. Wildlife habitats in Tucson: a strategy for conservation. Univ. Ariz. School Renewable Nat. Resour., Tucson.
Stromberg, J.C., J.A. Tress, S.D. Wilkins and S.D. Clark. 1992. Response of velvet mesquite to groundwater decline. J. Arid Envir. 23:45-58.
Temple, S.A., and B. Wilcox. 1986. Predicting effects of habitat patchiness and fragmentation. In: J. Verner, M.L. Morrison, and C.J. Ralph (eds.), Wildlife 2000: Modeling habitat relationships of terrestrial vertebrates, pp. 261-262. University of Wisconsin Press, Madison, WI.
Zar, J.H. 1984. Biostatistical analysis. Prentice-Hall, Inc., Englewood Cliffs, NJ.

Appendix 1. Species list and habitat where found for study sites along Cave Crek Wash.

			Habitat		
	Scientific name	Common name	Urban	Margin	Rural
Tree	*Cercidium floridum*	Blue Palo Verde	X	X	X
	Chilopsis linearis	Desert Willow	X	X	X
	Olneya tesota	Desert Ironwood		X	X
	Parkinsonia aculeata	Mexican Palo Verde	X		
	Prosopis velutina	Honey Mesquite		X	X
	Tamarix chinensis	Tamarisk	X	X	
	Washingtonia robusta	Mexican Fan Palm	X		
Shrub					
	Acacia greggii	Catclaw Acacia	X	X	X
	Ambrosia ambrosioides	Canyon Ragweed		X	X
	Ambrosia deltoidea	Bur Sage	X	X	X
	Ambrosia eriocentra	Ragweed	X	X	
	Atriplex canescens	Four-wing Saltbush	X		
	Baccharis sarathroides	Desert Broom	X	X	X
	Bebbia juncea	Bebbia	X	X	X
	Celtis pallida	Desert Hackberry		X	X
	Cassia spp.	Cassia	X		
	Dasylirion wheeleri	Sotol	X		
	Encelia farinosa	Brittlebush	X	X	X
	Ephedra trifurca	Mormon Tea		X	X
	Erigonum fasiculatum	Desert Buckwheat			X
	Haplopappus spp.	Haplopappus	X	X	
	Hymenoclea salsola	Burro-brush	X	X	X
	Larrea tridentata	Creosote	X	X	X
	Lycium andersonii	Desert Wolfberry		X	X
	Lycium exsertum	Wolfberry	X	X	X
	Sarcostemma cynanchoides	Climbing Milkweed		X	X
	Simmondsia chinensis	Jojoba	X	X	
	Vauguelinia californica	Arizona Rosewood			X
	Ziziphus obtusifolia	Graythorn	X	X	
Cacti					
	Carnegia gigantea	Saguaro		X	X
	Ferocactus acanthoides	Red Barrel Cactus		X	X
	Ferocactus wislizenii	Arizona Barrel Cactus			X
	Echinocereus engelmannii	Hedgehog Cactus		X	X
	Opuntia bigelovii	Teddybear Cactus			X

Appendix 1. Continued.

Scientific name	Common name	Habitat urban	margin	rural
Opuntia fulgida	Jumping Cholla			X
Opuntia leptocaulis	Christmas Cholla		X	X
Opuntia phaeacantha	Pricklypear Cactus			X
Dicot herb				
Allionia incarnata	Trailing-four O'clock			X
Bowlesia incana			X	
Brassica eruca	Golden Rocket		X	X
Centaurea solstitialis	5-star Thistle	X	X	
Datura wrightii	Jimson Weed	X	X	X
Eriogonum deflexum	Flatcrown Eriogonum	X		
Erodium cicutarium	Fillaree	X	X	X
Euphorbia albomarginata	Spotted Euphorbia	X		
Gilia achilleaefolia	California Gilia		X	
Lepidium montanum	Peppergrass		X	
Medicago polymorpha	Bur Clover	X		
Meliotus albus	White-seed Clover	X		
Nicotiana trigonophylla	Desert Tobacco	X	X	X
Nicotiana glauca	Tree Tobacco		X	
Proboscidea altheaefolia	Golden Devil's-Claw			X
Raphanus sativus	Wild Radish	X	X	X
Salsola kali	Tumbleweed	X		
Sphaeralcea ambigua	Desert Mallow	X	X	X
Tiderstromia lanuginosa	Tiderstromia		X	
Veronica peregrina	Speedwell	X		
Xanthusa spp.	Cockle-bur	X		
Monocots				
Bromus tectorum	Downy Chess	X	X	X
Bromus carinatus	California Brome			X
Bromus diandra	Rip-gut brome	X		
Bromus rubens	Red brome	X	X	X
Cynodon dactylon	Bermuda Grass	X		
Pennisetum setaceum	Fountain Grass	X		
Phalaris minor	Canary Grass	X	X	X
Polypogon monspeliensis	Rabbit-foot Grass			X
Schismus barbatus	Schismus	X	X	X
Typha augustifolia	Cattail	X	X	

EXOTIC HERBACEOUS SPECIES IN ARIZONA'S RIPARIAN ECOSYSTEMS

J. C. Stromberg, L. Gengarelly, and B. F. Rogers
Center for Environmental Studies, Arizona State University, Tempe, AZ 85287-3211 USA

Abstract

Herbaceous vegetation was sampled to determine whether cover and richness of exotic species differed between patch types located along floodplain moisture-disturbance gradients. Data were collected during three seasons at three rivers within each of two regions of Arizona. Of 199 herbaceous species sampled along the six rivers, 23% were exotic and 77% were native. During summer and fall, exotic cover was significantly higher on stream banks (the wettest and most frequently inundated and flood-scoured patch type) than on the drier, less frequently flooded *Populus fremontii-Salix gooddingii* (Fremont cottonwood-Goodding willow) and *Prosopis velutina* (velvet mesquite) terraces. Stream banks had greater than 55% relative exotic cover in all seasons and both regions. These findings support the theories that exotics are more abundant in early than late seral stages, and more abundant in mesic than more xeric habitats. In spring, however, exotic cover was significantly higher in mesquite woodlands (85%) than in the two other patch types (<25%), in the Sonoran Desert region. Most of the spring-season exotic cover was composed of cool-season annuals (e.g., *Bromus rubens, Hordeum murinum*) that germinate after winter/spring rains. Dominant exotics on stream banks included a mix of annuals (e.g., *Polypogon monspeliensis*) and warm-season, clonal perennials (e.g., *Cynodon dactylon, Sorghum halepense*). There is a need for research on effects of these and other widespread herbaceous exotics on structure, composition, and function of riparian ecosystems of the arid Southwest.

Introduction

Habitats in which the endogenous disturbance regimes have been altered are among the most susceptible to biological invasions (Fox and Fox 1986; McIntyre *et al.* 1988). In riparian ecosystems, floods are natural disturbances that create an abundance of temporal and spatial niches within floodplains by triggering successional processes and contributing to the formation of habitat mosaics (patch types). Flood flows interact with vegetation to produce spatial gradients of depth to groundwater; inundation frequency; soil texture, moisture, and salinity; and light quantity, all of which allow for a high diversity of native plant species in riparian ecosystems (Naiman *et al.* 1993). In many rivers of the US Southwest, however, the flooding regime has been altered by river regulation, watershed degradation, and/or stream dewatering. Additionally, riparian zones have been used for agriculture, livestock grazing, and urban development, all of which constitute forms of exogenous disturbance (Baker *et al.* 1993). Riparian zones also are frequently used as travel corridors for roads and railways, thereby providing frequent opportunities for accidental intro-

Plant Invasions: Studies from North America and Europe, pp. 45–57
edited by J.H. Brock, M. Wade, P. Pyšek and D. Green

duction of exotics (Frenkel 1977). As a consequence, riparian ecosystems frequently contain many non-native species (Nilsson *et al.* 1989; du Preez and Venter 1990; DeCamps 1993; Wolden *et al.* 1994).

In the American Southwest, *Tamarix chinensis* (salt cedar and related species) and *Elaeagnus angustifolia* (Russian olive) are the most widely distributed woody exotics (Harris 1966; Olson and Knopf 1986; Brock 1994). In Arizona, *Tamarix* spp. stands cover about five times as much area as do native forests of *Populus fremontii* (Fremont cottonwood) and *Salix gooddingii* (Goodding willow), due in part to river damming and salinization of waters by agricultural activities (Valencia *et al.* 1993). Once established, *Tamarix* spp. can alter understory plant diversity and ecosystem properties by increasing fire frequency, sedimentation rate, and soil salinity (e.g. Griffin *et al.* 1989; Brock 1994). Riparian ecosystems also have many exotic herbaceous species, but little is known of their distribution patterns or ecosystem effects.

Several factors may influence the distribution of herbaceous exotic species in riparian ecosystems. For example, exotics decline with successional status in some ecosystems (Rejmanek 1989), and thus might have greatest cover on the most frequently inundated sites. McIntyre *et al.* (1988), in contrast, reported that high levels of endogenous disturbance help to confer resistance to invasion by exotics, and others suggest that flooding may be a natural mechanism that restores native species to sites floristically altered by grazing (Chaneton *et. al.* 1988). Exotic species also have been reported to be more abundant in mesic parts of ecosystems than in xeric or hydric parts (Rejmanek 1989). The primary objective of this study was to determine whether cover and richness of herbaceous exotics varied among riparian patch types located along moisture-disturbance gradients. A second objective was to identify native plant species with similar distribution patterns to invasive exotics.

Study areas

Study areas included six alluvial, perennial, free-flowing rivers, all of which have been grazed by livestock in the recent past (Fig. 1). Three were in the Sonoran Desert region of central Arizona (Hassayampa River, Date Creek, and Santa Maria River), and three were in the semidesert grassland region of southern Arizona (San Pedro River, Sonoita Creek, and Santa Cruz River). The central Arizona river sites were at low elevation (370–880 m) and had low average annual rainfall (20 to 27 cm/year), about half of which falls in the winter rainy season and half in the summer (July-September) rainy season. The southern Arizona sites were at higher elevation (1000 to 1220 m) and had higher rainfall (37 to 43 cm/yr), about two-thirds of which falls in summer.

The central Arizona sites were vegetated by Sonoran Riparian Deciduous Forests (Brown 1982), and supported *P. fremontii* and *S. gooddingii* on lower floodplains and *P. velutina* on higher floodplain terraces. *T. chinensis* formed a small component of the overstory vegetation. Shrubs intermixed with the forests included *Baccharis salicifolia* (seepwillow) along river banks; *Hymenoclea monogyra* (burro brush) on mid-elevation floodplains; and *Zizyphus obtusifolia* (graythorn), *Acacia greggii* (catclaw), and *Lycium andersonii* (wolfberry) on high floodplain terraces. The southern Arizona sites were vegetated by the *P. fremontii-S. gooddingii* series of the Warm Temperate Interior Riparian Deciduous Forest (Brown 1982), and also supported

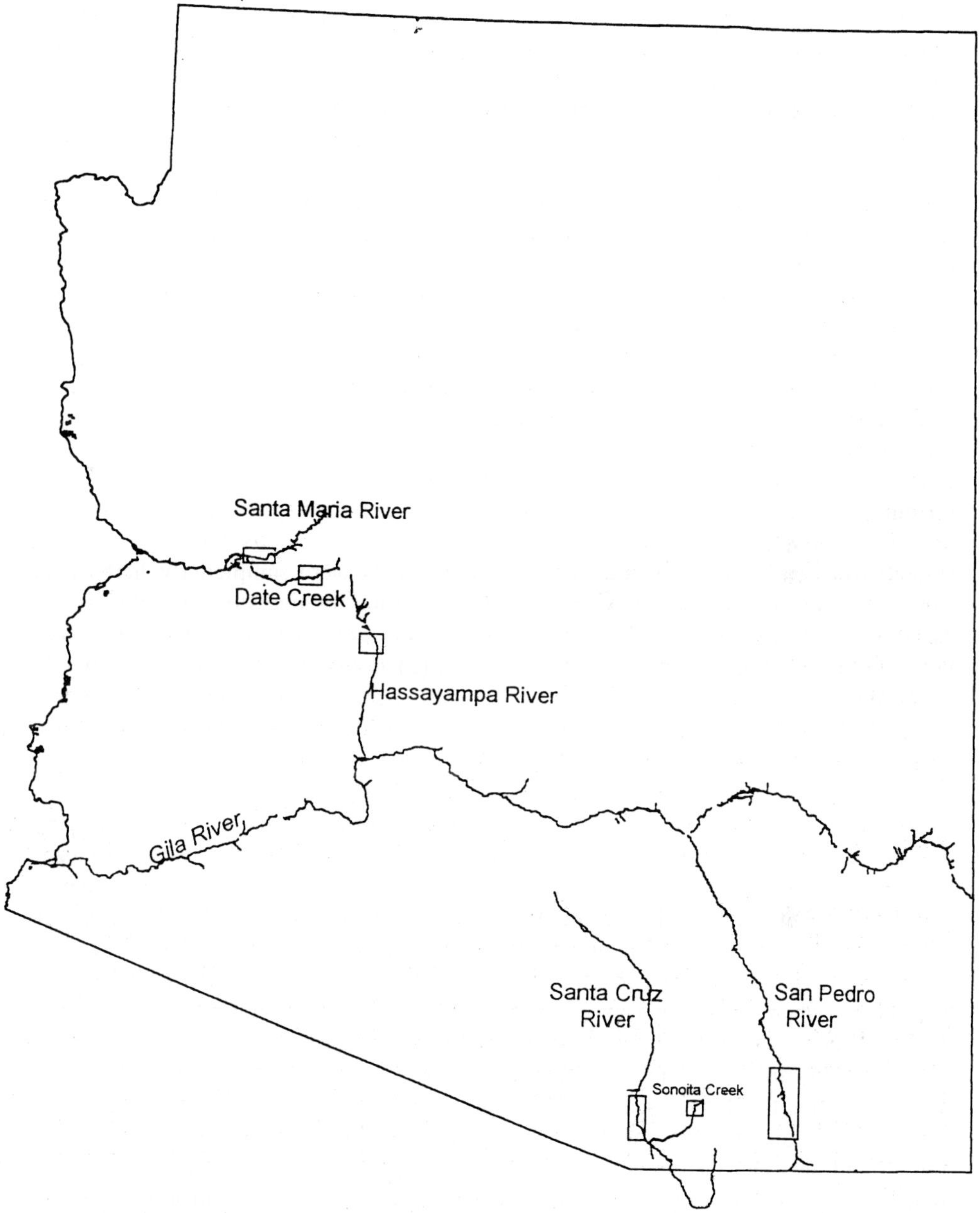

Fig. 1. Location of study areas in Arizona.

broadleaf trees including *Juglans major* (Arizona walnut), *Fraxinus velutina* (Arizona ash), and *Celtis reticulata* (netleaf hackberry). *Chrysothamnus nauseosus* (rabbit-brush) co-occurred with *H. monogyra* on mid-elevation floodplain terraces. *Sporobolus wrightii* (sacaton) grasslands were intermixed with woodlands on high floodplain terraces.

The Hassayampa River study area (Fig. 1) was managed as a preserve by The Arizona Nature Conservancy (Stromberg *et al.* 1993a). Land uses prior to 1987 included livestock grazing, orchard cultivation, and recreational use. The Date Creek study area was on Date Creek Ranch, and was grazed by cattle during winter. The Santa Maria study area was on Bureau of Land Management lands and was grazed by feral burros and trespass livestock. The San Pedro study area encompassed a 50 km river reach managed by the Bureau of Land Management as a Riparian National Conservation Area (Stromberg *et al.* 1996). Livestock, sand and gravel mining and many other land uses have been excluded since 1987. The Sonoita Creek study area was in The Arizona Nature Conservancy's Patagonia-Sonoita Preserve, and was grazed and farmed prior to 1966 (Rogers 1994). The Santa Cruz River study area was downstream of the Nogales International Wastewater Treatment plant, which discharges effluent into the Santa Cruz river channel (Stromberg *et al.* 1993b). The Santa Cruz study area was grazed by trespass livestock.

Methods

At the six rivers, 40 to 60 randomly placed quadrats were sampled for herbaceous cover, by species, within three patch types differing in overstory vegetation and landform type, including: (1) stream banks and within-channel sand bars (i.e., Brown's [1982] Sonoran Interior Strand zone); (2) floodplain terraces supporting 15 to 40 yr *P. fremontii-S. gooddingii* forests; and (3) floodplain terraces supporting *P. velutina* woodlands. The three patch types occur along a gradient from high soil moisture, shallow groundwater, and frequent inundation; to low soil moisture, deep groundwater, and infrequent inundation (Table 1; Stromberg *et al.* 1993a; Stromberg *et al.* 1996). Generally, the Sonoran Interior strand forms a narrow, continuous strip bordering the active channel. *P. velutina* woodlands form a wide, continuous band at the floodplain perimeter, and often occupy greater area than any other riparian patch type. *Populus-Salix* forests lie between the stream bank and *P. velutina* woodland zone, and often occur in non-continuous patches (Stromberg *et al.* 1997).

Plot size was 1 × 1 m except in the San Pedro and Sonoita Creek *P. velutina* woodlands where plot size was 1 × 0.5 m during the fall sampling period. Herbaceous cover was sampled (visual estimation) in three seasons: spring (March-April), summer (May-July), and fall (September-October), during 1990 to 1994. Plot locations were not fixed over seasons. The spring sampling period followed the winter rainy season, and summer and fall sampling periods represent pre- and post-summer rain conditions. During the study, Pacific frontal storms caused regional flooding in winters of 1991 and 1993. For example, the Hassayampa River had a 10-yr recurrence flood in March, 1991 and a 25-yr recurrence flood in January, 1993. Values for absolute cover and richness of native and exotic species were compared among patch types using one-way analysis of variance followed by Scheffe's multiple comparison test, for each season within each region (SPSS PC software). Relative values for exotic cover and richness were calculated from mean site values (i.e., mean cover of exotics as a percentage of mean total cover).

Species names follow Lehr (1978) and recent treatises published in the Journal of the Arizona-Nevada Academy of Sciences. Determination of exotic or native status

Table 1. Abiotic characteristics of three riparian patch types along rivers in the Sonoran Desert (Hassayampa River) and semidesert grassland region (San Pedro River) of Arizona.

	Depth to groundwater (m)	Inundation frequency (years)	Soil texture Silt+clay (%)	Soil organic matter (%)	Soil nitrate (ppm)
Hassayampa River					
Stream bank	0.3	1	–	0.9	12
Populus-Salix forest	2.2	3	–	1.9	17
Prosopis woodland	3.1	8	–	2.2	89
San Pedro River					
Stream bank	0.1	1	22	0.7	5
Populus-Salix forest	1.9	5	45	1.3	6
Prosopis woodland	5.2	>100	50	2.1	14

was based on review of several floras including Kearney and Peebles (1960), Correll and Correll (1972), and Hickman (1993).

Results

Of 199 herbaceous species sampled at the six rivers, 23% (45) were exotic and 77% were native. Most of the exotics (67%) had an annual life cycle (Table 2). Larger percentages of the exotics were classified as facultative wetland, facultative, and facultative upland species, and fewer as obligate upland or wetland species, in comparison to native species (Table 3).

Stream bank patch type

Exotic cover was significantly higher on stream banks (the most frequently inundated and wettest patch type) than on *Populus-Salix* and *P. velutina* terraces, during summer and fall in both regions (ANOVA, $P < 0.05$) (Fig. 2). Relative exotic cover was higher on stream banks in summer and fall at the southern Arizona rivers, but did not vary greatly among patch types in central Arizona. Exotic species richness was significantly higher on stream banks than in the terrace patch types during spring and fall in southern Arizona (ANOVA, $P < 0.05$) (Fig. 3).

In all seasons and regions, stream banks had greater than 55% relative exotic cover. The most prevalent exotic species in the wettest portion of this patch type was *Nasturtium officinale* (water-cress), a cool-season perennial recorded at all six rivers. Native species in this habitat included *Hydrocotyle ranunculoides* and *H. verticillata* (water pennyworts), *Mimulus guttatus* (monkey flower), *Paspalum distichum* (knot grass), and *Veronica anagallis-aquatica* (water speedwell). On sandbars within the channel, the annual *Polypogon monspeliensis* (rabbit's foot grass) was abundant at all rivers during summer. Native species in this same habitat, where water was at the soil surface or a few centimeters below, included several perennial monocots including *Cyperus odoratus* (flat sedge), *Eleocharis montevidensis* (spike rush), *Juncus torreyi*, *J. articulatus, J. xiphioides* (rushes), *Scirpus americanus* (bulrush), and *Typha domin-*

Table 2. Life-history attributes and wetland indicator status (Reed, 1988) of exotic herbaceous species in riparian zones of six perennial rivers in Arizona. (Ann = annual; Bien = biennial; Per = perennial); (S = sexual [seed], A = asexual spread [e.g., rhizomes]).

Family	Species	Life cycle	Reproductive method	Wetland status
Asteraceae	*Centaurea melitensis*	Ann	S	Obligate upland
	Helianthus annuus	Ann	S	Facultative
	Gnaphalium luteo-album	Ann	S	Facultative
	Lactuca serriola	Ann	S	Facultative
	Sonchus asper	Ann	S	Facultative wetland
	Sonchus oleraceus	Ann	S	Obligate upland
Brassicaceae	*Brassica nigra*	Ann	S	Obligate upland
	Capsella bursa-pastoris	Ann	S	Facultative
	Nasturtium officinale	Per	S&A	Obligate wetland
	Sisymbrum irio	Ann	S	Obligate upland
Chenopodiaceae	*Chenopodium album*	Ann	S	Facultative
	Chenopodium ambrosioides	Ann/Per	S	Facultative
	Chenopodium murale	Ann	S	Obligate upland
	Salsola iberica	Ann	S	Facultative upland
Convolvulaceae	*Ipomoea purpurea*	Ann	S	Obligate upland
Fabaceae	*Melilotus albus*	Bien	S	Facultative upland
	Melilotus officinalis	Bien	S	Facultative upland
Geraniaceae	*Erodium cicutarium*	Ann	S	Obligate upland
Lamiaceae	*Lamium amplexicaule*	Ann	S	Obligate upland
	Marrubium vulgare	Per	S	Facultative
Plantaginaceae	*Plantago major*	Per	S	Facultative wetland
Poaceae	*Avena fatua*	Ann	S	Obligate upland
	Bromus catharticus	Ann	S	Obligate upland
	Bromus diandrus	Ann	S	Obligate upland
	Bromus rubens	Ann	S	Obligate upland
	Bromus sterilis	Ann	S	Obligate upland
	Cortaderia selloana	Per	S	Obligate upland
	Cynodon dactylon	Per	S&A	Facultative upland
	Digitaria sanguinalis	Ann	S	Facultative upland
	Echinochloa crus-galli	Ann	S	Facultative wetland
	Eragrostis cilianensis	Ann	S	Facultative upland
	Eragrostis lehmanniana	Per	S	Obligate upland
	Hordeum murinum	Ann	S	Obligate upland
	Paspalum dilatatum	Per	S&A	Facultative
	Polypogon interruptus	Ann	S	Obligate wetland
	Polypogon monspeliensis	Ann	S	Facultative wetland
	Polypogon viridis	Per	S	Facultative wetland
	Schismus barbatus	Ann	S	Obligate upland
	Sorghum halapense	Per	S&A	Facultative upland
Polygonaceae	*Polygonum aviculare*	Ann	S	Facultative wetland
	Polygonum lapathifolium	Ann	S	Obligate upland
	Rumex crispus	Per	S	Facultative wetland
	Rumex obtusifolius	Per	S	Facultative wetland
Rubiaceae	*Galium aparine*	Ann	S	Facultative upland
Scrophulariaceae	*Verbascum thapsus*	Bien	S	Obligate upland

Table 3. Wetland indicator status (Reed 1988) of native and exotic herbaceous species in the riparian zones of six Arizona rivers (pooled values).

	Total taxa	Native taxa	Exotic taxa
Obligate wetland	21	19 (12%)	2 (4%)
Facultative wetland	23	15 (10%)	8 (18%)
Facultative	26	18 (12%)	8 (18%)
Facultative upland	16	8 (5%)	8 (18%)
Obligate upland	113	94 (61%)	19 (42%)
TOTAL	199	154	45

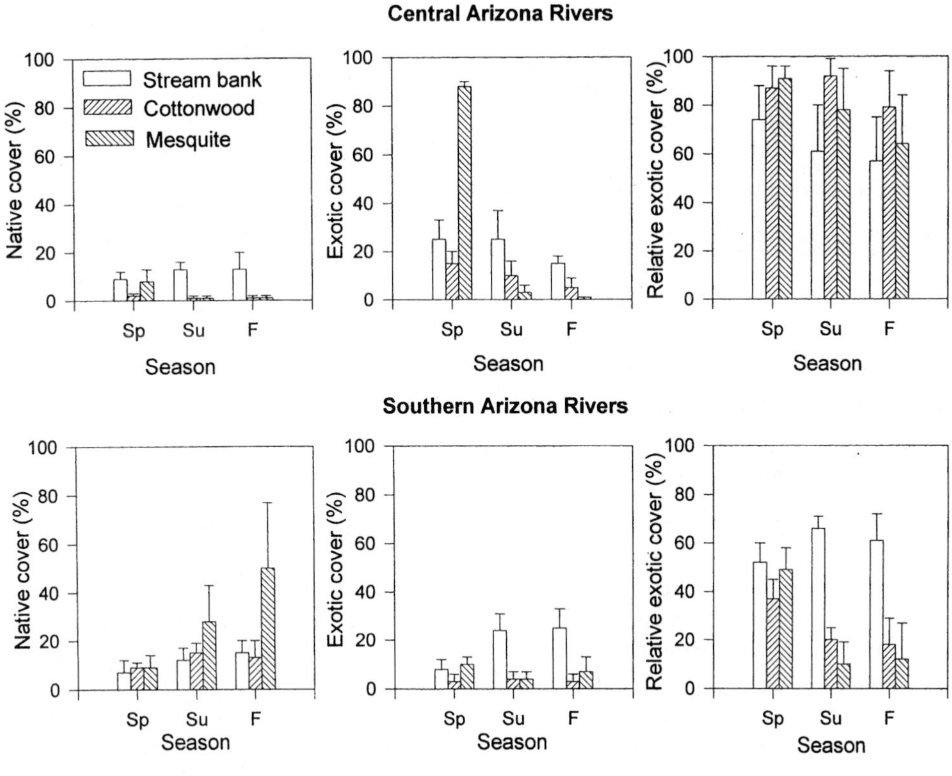

Fig. 2. Cover of native and exotic herbaceous species, and relative cover of exotic species, in three riparian patch types during three seasons, in two regions of Arizona. Values shown are means (with standard error bars) for three rivers per region (Sp=spring; Su=summer; F=fall).

gensis (tropical cattail). Others included *Castilleja cruenta* (Indian paint brush) and *Leptochloa uninervia* and *L. fascicularis* (sprangletop grasses).

Cynodon dactylon (bermuda grass), a rhizomatous grass, and *Melilotus albus* (white sweet clover), a biennial forb, were common exotics on stream banks and low floodplain terraces at most rivers. The perennial *Sorghum halapense* (Johnson grass)

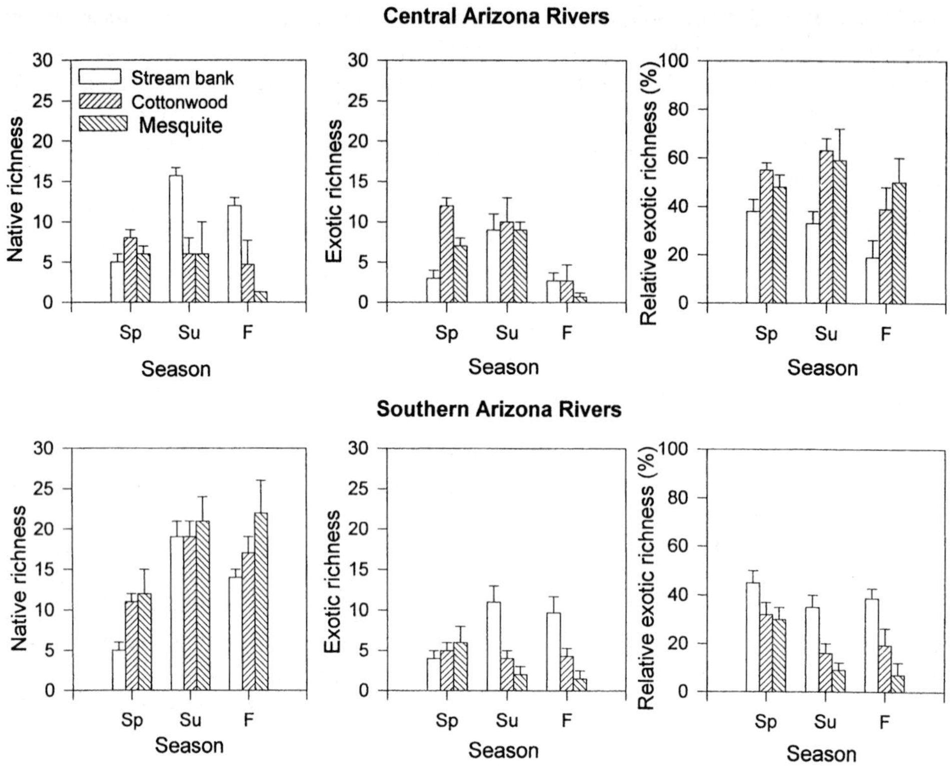

Fig. 3. Richness of native and exotic herbaceous species (species number per 50 m^2), and relative richness of exotic species, in three riparian patch types during three seasons, in two regions of Arizona. Values shown are means (with standard error bars) for three rivers per region (Sp=spring; Su=summer; F=fall).

was abundant along stream banks in southern Arizona. Native species with similar distribution patterns to these three exotics included a mixture of annuals and perennials, including *Centaurium calycosum* (Buckley's centaury), *Calibrachoa parviflora* (wild petunia), *Epilobium californicum* (willow weed), *Equisetum laevigatum* (horsetail), *Muhlenbergia rigens* (deer grass), *Polygonum punctatum* (water smart-weed), and *Xanthium strumarium* (cocklebur).

Prosopis velutina woodlands

In spring in central Arizona, exotic cover was significantly higher in *P. velutina* woodlands (>85% cover) than in the other two patch types (<25% cover) (ANOVA, P <0.05) (Fig. 2). Relative exotic cover was high (>70%) in all three patch types in spring in both regions. Exotic winter annuals were the dominant species in the *P. velutina* understories in spring in central Arizona. These included *Bromus catharticus* (rescue brome), *Bromus diandrus* (ripgut brome), *Bromus rubens* (red brome), *Erodium cicutarium* (filaree), *Hordeum murinum* (foxtail barley), *Schismus barbatus* (Meditteranean grass), and *Sisymbrium irio* (London rocket). Native winter annuals, including *Amsinckia intermedia* (coast fiddleneck), *Bowlesia incana* (hairy bowlesia), and *Parietaria hespera* (pellitory), were present in low abundance. The understory of the central Arizona *P. velutina* woodlands in summer was sparse. Summer exotics included *Marrubium vulgare* (horehound), an evergreen perennial, *Centaurea melitensis* (Malta star thistle), and *Salsola iberica* (Russian thistle). Native summer species in central Arizona included *Nicotiana trigonophylla* (desert tobacco) and *Abutilon parvulum* (small-leaved abutilon). Many native species were present in summer and fall in the southern Arizona *P. velutina* woodlands, including *Acalypha neomexicana* (three-seeded mercury), *Bidens leptocephala* (bur marigold), *Datura meteloides* (sacred datura), *Ipomoea hederacea* (morning-glory vine), and S. *wrightii.*

Populus fremontii-Salix gooddingii forests

Native and exotic herbaceous cover were generally sparse in the *Populus-Salix* patch type (Fig. 2). In central Arizona, greatest cover of exotics in this patch type occurred in spring. Cool-season exotics included several annual grasses (e.g., *H. murinum, B. rubens,* and *B. diandrus*). Native cool-season annuals included *A. intermedia*, *B. incana*, and *P. hespera*. *Cymodon dactylon* was the most abundant exotic in this patch type in summer, and was present at all six rivers. Other frequently encountered exotics in summer included *Lactuca serriola* (prickly lettuce), *Sonchus oleraceus* (sow thistle), and *S. iberica*. Native summer annuals included *Conyza canadensis* (horseweed), *Heterotheca psammophila* (camphor weed), and *X. strumarium.* Native perennials included *Ambrosia psilostachya* (western ragweed), *Anemopsis californicum* (yerba-mansa), *D. meteloides, Elymus canadensis* (Canada wild-rye), *Mirabilis longiflora* (sweet four o'clock), and *N. trigonophylla*.

Discussion

Patterns of variation between patch types at six Arizona rivers provide some support for the theories that exotic species have greater abundance in early seral than late seral patch types, and on mesic than on more xeric sites. The one notable exception to the trend for greatest exotic cover on stream banks, occurred in spring at the Sonoran Desert rivers. Many of the exotic winter annuals (e.g., *B. rubens* and *S. barbatus, H. murinum*, *S. irio)* with high cover in the nitrogen-rich *Prosopis* woodlands are native to the European Mediterranean region, and are pre-adapted to the temperate, winter-rain conditions characteristic of spring in the Sonoran Desert region. Many also are more drought tolerant than native winter annuals, perhaps contributing to their abundance in the high terrace woodlands (Beatley 1966, 1974; Davison 1971; Wu and Jain 1979). Additionally, *B. rubens* germinates earlier than many native winter annuals because it has less specific moisture and temperature germination requirements, thus qualifying it as a "gap grabber" that pre-empts available germination space (Newsome and Noble 1986). Further study is needed to determine if the high cover of the exotic winter annuals is altering ecosystem processes (Bock *et al.* 1986; Vitousek 1986; Shafroth *et al.* 1994), or having no discernable effect (Simberloff 1981). Dense cover of the exotics may reduce cover and diversity of native winter annuals (e.g., *A. intermedia, B. incana*) as well as of warm season species by forming a dense surface litter layer (e.g., similar to *Bromus tectorum* in the Great Basin Desert, Billings 1990). High densities of *B. rubens* also may cause local declines in rodent populations (Rowland and Turner 1964) and increase in fire spread (Beatley 1966).

Many of the dominant exotics on stream banks were perennials with the capacity for sexual and asexual spread (e.g., *C. dactylon, N. officinale, S. halepense*), traits common to many invasive exotics (Baker 1965; Holm *et al.* 1977; Crawley 1987). *Cynodon dactylon* has high survivorship of floods, drought, and salinity, and can maintain itself for long time periods through rhizomatous spread (Horowitz 1972; Devitt *et al.* 1993). Invasive species with such traits are considered as "survivors" (long-lived individuals resistant to most causes of mortality) (Newsome and Noble 1986). *Melilotus alba* and *M. officinalis,* also dominant stream bank exotics, may gain competitive advantage over some native stream bank species because of the nitrogen supplied by their symbiotic nitrogen-fixing bacteria. *Cynodon dactylon* and *P. monspeliensis* also form associations with nitrogen-fixing bacteria in the rhizosphere which may increase their growth rate in young, infertile floodplain soils (Malik *et al.* 1991). With regard to ecological consequences, *C. dactylon* and other exotics have been variously reported to reduce growth of woody seedlings and other species by reducing soil moisture, soil nutrients, or light levels, or through allelopathy (Horowitz and Friedman 1971; Cohn *et al.* 1989; Meissner *et al.* 1989); however native species also may cause these effects. Similarly, although *C. dactylon* may increase substrate stability during floods, it is unclear to what extent it is replacing the functions of native rhizomatous species *(e.g., Distichlis spicata, P. distichum*) that may have been locally extirpated. Studies are needed that compare the functions of herbaceous exotics in Southwest USA riparian ecosystems with those of native species that have similar distribution patterns.

Much time, effort and money have been spent on mechanical, chemical, and biological control of invasive species (McKnight 1993). Exotics also may be controlled by identifying and restoring processes and conditions that favor native species. For example, livestock grazing may have contributed to the invasion of annual grasses into Southwestern desert ecosystems adapted to only moderate grazing pressure (Mack 1986; Burgess *et al.* 1991; Brooks 1995). In this study, all but one site was ungrazed at present, but past histories of grazing and agriculture similarly may have introduced and facilitated the establishment of grazing tolerant exotic species including *C. dactylon* (Belsky 1986; Rucks 1984) and *N. officinale* (Rejmankova 1992). Rogers (1994) reported that perennial native grasses were more abundant in *P. velutina* understories from which cattle had been excluded, although exotics were still present at sites with the longest (25 yr) grazing-free period. In any event, the need is clear for additional research on environmental factors and management practices that favor native species over exotics in riparian ecosystems of the arid U. S. Southwest.

Acknowledgements

Thanks are due to Dawn Hagerman, Leigh Hedrick, Dr. John Reeder, Charlotte Reeder, and Joelle Don de Ville for assistance with data collection and species identification. Review comments from Barb Tellman, John Randall, and others are greatly appreciated. We are grateful to the Arizona Nature Conservancy, Bureau of Land Management, Date Creek Ranch, and others for allowing access to study sites.

References

Baker, H. G. 1965. Characteristics and modes of origin of weeds. In: The genetics of colonizing species, H. G. Baker and G. L. Stebbins (eds.), Academic Press, New York.

Baker, R. G., D. P. Schwert, E. A. Bettis III, and C. A. Chumbley. 1993. Impact of Euro-American settlement on a riparian landscape in northeast Iowa, midwestern USA: an integrated approach based on historical evidence, floodplain sediments, fossil pollen, plant macrofossils and insects. The Holocene 3:314-323.

Beatley, J. C. 1974. Phenological events and their environmental triggers in Mojave desert ecosystems. Ecology 55:856-863.

Beatley, J. C. 1966. Ecological status of introduced brome grasses (*Bromus* spp.) in desert vegetation of southern Nevada. Ecology 47:548-554.

Belsky, A. J. 1986. Revegetation of artificial disturbances in grasslands along the Serengeti National Park, Tanzania: colonization of grazed and ungrazed plots. Journal of Ecology 74:419-437.

Billings, W. D. 1990. *Bromus tectorum*, a biotic cause of ecosystem impoverishment in the Great Basin. In: G. M. Woodell (ed.), The earth in transition: patterns and processes of biotic impoverishment. pp. 301–322. Cambridge University Press, Cambridge.

Bock, C. E., J. H. Bock, K. L. Jepson and J. C. Ortega. 1986. Ecological effects of planting African lovegrasses in Arizona. Natural Geographic Research 2:456-463.

Brock, J. H. 1994. *Tamarix* spp. (salt cedar), an invasive exotic woody plant in arid and semi-arid riparian habitats of Western USA. Pages 27-44 in L. C. de Waal, L. E. Child, P. M. Wade and J. H. Brock (eds.), Ecology and management of invasive riverside plants. John Wiley and Sons, Chichester.

Brooks, M. L. 1995. Benefits of protective fencing to plant and rodent communities of the Western Mojave Desert, California. Environmental Management 19:65-74.

Brown, D. E. (ed.). 1982. Biotic communities of the American Southwest-United States and Mexico. Desert Plants 4:1-342.

Burgess, T. L., J. E. Bowers and R. M. Turner. 1991. Exotic plants at the Desert Laboratory, Tucson, Arizona. Madrono 38:96-114.

Chaneton, E. J., J. M. Facelli, and R. J. C. Leon. 1988. Floristic change induced by flooding on grazed and ungrazed lowland grasslands in Argentina. Journal of Range Management 41:495-499.

Cohn, E. J., O. W. Van Auken and J. K. Bush. 1989. Competitive interactions between *Cynodon dactylon* and *Acacia smallii* seedlings at different nutrient levels. American Midland Naturalist 121:265-272.

Correll, D. S. and H. B. Correll. 1972. Aquatic and wetland plants of Southwestern United States. Environmental Protection Agency, U. S. Government Printing Office, Washington DC.

Cox, J. R., H. L. Morton, T. N. Johnson Jr., G. L. Jordan, S. C. Martin, and L. C. Fierro. 1984. Vegetation restoration in the Chihuahuan and Sonoran Deserts of North America. Rangelands 6:112-116.

Crawley, M. J. 1987. What makes a community invasible? In: A. J. Gray, M. J. Crawley, and P. J. Edwards (eds.), Colonization, succession and stability. pp. 429–453. Blackwell, Oxford.

Davison, A. W. 1971. The ecology of *Hordeum murinum* L. II. The ruderal habit. Journal of Ecology 59:493-506.

DeCamps, H. 1993. River margins and environmental change. Ecological Applications 3:441-445.

Devitt, D. A., D. C. Bowman and P. J. Schulte. 1993. Response of *Cynodon dactylon* to prolonged water deficits under saline conditions. Plant and Soil 148:239-251.

Du Preez, P. J., and H. J. T. Venter. 1990. The phytosociology of the woody vegetation in the southern part of the Vredefort Dome area. Part 1: communities of the plains, riverbanks, and islands. South African Journal of Botany 56:631-636.

Fox, M. D. and B. J. Fox. 1986. The susceptibility of natural communities to invasion. In: R. H. Groves and J. J. Burdon (eds.), Ecology of biological invasions. pp. 57–66. Cambridge University Press, New York.

Frenkel, R. E. 1977. Ruderal vegetation along some California roadsides. University of California Publications in Geography 20:1-163.

Griffin, G. F., D. M. Stafford Smith, S. R. Morton, G. E. Allan, and K. A. Masters. 1989. Status and implications of the invasion of tamarisk (*Tamarix aphylla*) on the Finke River, Northern Territory, Australia. Journal of Environmental Management 29:297-315.

Harris, D. R. 1966. Recent plant invasions in the arid and semi-arid Southwest of the United States.

Annals of the Association of American Geographers 56:408-422.

Hickman, J. C. (ed.). 1993. The Jepson manual: higher plants of California. University of California Press, Berkeley.

Holm, L. G., D. L. Plucknett, J. V. Pancho, and J. P. Herberger. 1977. The world's worst weeds, distribution and biology. University Press, Hawaii.

Horowitz, M. 1972. Effects of dessication and submergence on the viability of rhizome fragments of Bermuda grass and Johnson grass and tubers of nutsedge. Israel Journal of Agricultural Research 22:215-220.

Horowitz, M. and T. Friedman. 1971. Biological activity of subterranean residues of *Cynodon dactylon* L., *Sorghum halapense* L. and *Cyperus rotundus* L. Weed Research 11:88-93.

Kearney, T. H. and R. H. Peebles. 1960. Arizona flora (with supplement). University of California Press, Berkeley, California.

Lehr, J. H. 1978. A catalogue of the flora of Arizona. Desert Botanical Garden, Phoenix, Arizona.

Mack, R. N. 1986. Plant invasion into the intermountain West. In: H. A Mooney and J. A. Drake (eds.), Ecology of biological invasions of North America and Hawaii. pp. 191–213. Springer-Verlag, New York.

Malik, K. A., B. Rakhshanda, G. Rasul, K. Mahmood and M. I.Sajjad. 1991. Associative N_2-fixation in plants growing in saline-sodic soils and its relative quantification based on ^{15}N natural abundance. Plant and Soil 137:67-74.

McIntyre, S., P. Ladiges and G. Adams. 1988. Plant species-richness and invasion by exotics in relation to disturbance of wetland communities on the Riverine Plain, NSW. Australian Journal of Ecology 13:361-373.

McKnight, N. N. 1993. Biological pollution: the control and impact of invasive exotic species. Indiana Academy of Science, Indianapolis.

Meissner, R., P. C. Nel and E. A. Boyers. 1989. Allelopathic effects of *Cynodon dactylon*-infected soil on early growth of selected crop species. Applied Plant Science 3:125-126.

Naiman, R. J., H. DeCamps and M. Pollock. 1993. The role of riparian corridors in maintaining regional biodiversity. Ecological Applications 3:209-212.

Newsome, A. E. and I. R. Noble. 1986. Ecological and physiological characters of invading species. In: R. H. Groves and J. J. Burdon (eds.), Ecology of biological invasions. pp. 1–20. Cambridge University Press, New York.

Nilsson, C., G. Grelsson, M. Johansson and U. Sperens. 1989. Patterns of species richness along river banks. Ecology 70:77-84.

Olson, T. E. and F. L. Knopf. 1986. Naturalization of Russian-olive in the western United States. Western Journal of Applied Forestry 1:65-69.

Reed, P. B., Jr. 1988. National list of plant species that occur in wetlands: Southwest (region 7). U.S. Fish and Wildlife Service Biological Report 88(26.7).

Rejmanek, M. 1989. Invasibility of plant communities. In: J. A. Drake, H. A. Mooney, D. diCastri, R. H. Groves, F. J. Kruger, M. Rejmanek and M. Williamson, (eds.), Biological invasions: a global perspective. pp. 369–388. John Wiley and Sons, Chichester.

Rejmankova, E. 1992. Ecology of creeping macrophytes with special reference to *Ludwigia peploides* (H. B. K.) Raven. Aquatic Botany 32:283-299.

Rogers, B. 1994. An ecological survey of the understory of southwestern mesquite bosques and the impacts of grazing on this habitat. M. S. Thesis, Arizona State University, Tempe.

Rowland, R. H. and F. B. Turner. 1964. Correlation of the local distribution of *Dipodomys microps* and *D. merriami* and of the annual grass *Bromus rubens*. Southwestern Naturalist 9:56-61.

Rucks, M. G. 1984. Composition and trend of riparian vegetation on five perennial streams in southeastern Arizona. In: R. E. Warner and K. M. Hendrix (eds.), California riparian systems: ecology, conservation, and productive management. pp. 97–107. University of California Press, Berkeley.

Shafroth, P. B., M. L. Scott, R. D. Laven. 1994. Establishment, sex structure, and breeding system of an exotic riparian willow, *Salix × rubens*. American Midland Naturalist 132:159-172.

Simberloff, D. 1981. Community effects of exotic species. In: M. H. Nitecki (ed.), Biotic crises in ecological and evolutionary time. pp. 53–83. Academic Press, New York.

Stromberg, J. C., B. D. Richter, D. T. Patten and L. G. Wolden. 1993a. Response of a Sonoran riparian forest to a 10-year return flood. Great Basin Naturalist 53(2):118-130.

Stromberg, J. C., M. R. Sommerfeld, D. T. Patten, J. Fry, C. Kramer, F. Amalfi and C. Christian. 1993b. Release of effluent into the Upper Santa Cruz River, Southern Arizona: Ecological considerations. In: K. D. Schmidt (ed.), Proceedings of the symposium on effluent use manage-

ment. pp. 81–92. American Water Resources Association, Tucson, Arizona.
Stromberg, J. C., R. Tiller, and B. Richter. 1996. Effects of groundwater decline on riparian vegetation of semiarid regions: the San Pedro, Arizona. Ecological Applications 6:113-131.
Stromberg, J. C., J. Fry, and D. T. Patten. 1997. Vegetation change after large floods in an alluvial, arid-land river. Wetlands 17:292-300.
Valencia, R. A., J. A. Wennerlund, R. A. Winstead, S. Woods, L. Riley, and E. Swanson. 1993. Arizona riparian inventory and mapping project. Arizona Game and Fish Department, Phoenix, Arizona.
Vitousek, P. M. 1986. Biological invasions and ecosystem properties: can species make a difference? In: H. A Mooney and J. A. Drake (eds.), Ecology of biological invasions of North America and Hawaii. pp. 163–176. Springer-Verlag, New York.
Wolden, L., J. C. Stromberg, and D. T. Patten. 1994. Flora and vegetation of the Hassayampa River Preserve, Maricopa County, Arizona. Journal of the Arizona-Nevada Academy of Science 28:76-111.
Wu, K. K. and S. K. Jain. 1979. Population regulation in *Bromus rubens* and *Bromus mollis:* life cycle components and competition. Oecologia 39:337-358.

THE INVASIVE NATURE OF *LEPIDIUM LATIFOLIUM*: A REVIEW

James A. Young, Debra E. Palmquist, and Shay O. Wotring
USDA, Agricultural Research Service, 920 Valley Road, Reno, NV, 89512, USA

Abstract

Lepidium latifolium has rapidly spread to wetland and riparian areas in all of the far western states of the United States. Once established in wetlands it follows irrigation structures to agricultural fields. This is an extremely competitive weed that creates nearly complete monocultures once it becomes established. The weed is adapted to halomorphic soils, but is not obligate to such sites. Economic losses through reduced forage quality are significant. The impact on critical wetland-riparian habitats from the invasion of this weed pest is to create a significant decline in biological diversity. Control and quarantine efforts have largely been unsuccessful. The tremendous reproductive potential of the mass of creeping rootstocks is a major hinderance to effective control. Biological suppression through application of integrated pest management may be the key to dealing with this weed pest.

Introduction

Lepidium latifolium L. is formally known in the United States as perennial pepperweed. Among farmers and weed workers in the field, it is often referred to as tall whitetop. This common name arose from the myriad of densely packed white flowers attached to small panicles forming the heads or tops when in bloom. *L. latifolium* grows from 1 to 2 m or more in height and is a member of the *Brassicaceae* (mustard family). It forms colonies that dominate entire fields and wetland areas.

A recently published survey of *L. latifolium* infestation in the western United States (Young *et al.* 1995) reported the weed in all states except Arizona. More recently, *L. latifolium* has been collected in Arizona. This creeping-rooted perennial is native to southeastern Europe and southwestern Asia. It probably has been introduced into the United States multiple times as a suspected contaminant of sugar beet (*Beta vulgaris* L.) seed imported from eastern Europe. Before World War II importing of sugar beet seed was common. One of the first infestations in California was traced to such seed imported to Yolo County in the Sacramento Valley (Robbins 1940). *L. latifolium* is now found in all counties of California except the far northwestern area which is a temperate coastal rain forest and the far southeastern Sonoran desert (Young *et al.* 1995).

The infestations of *L. latifolium* in California are far from continuous. Characteristically, infestations first occur in riparian and wetland areas. Formerly these areas were often referred to by agriculturalists as "wastelands". Now that the ecological

Plant Invasions: Studies from North America and Europe, pp. 59–68
edited by J.H. Brock, M. Wade, P. Pyšek and D. Green

role of such areas in maintaining biological diversity is better understood, these wetland-riparian areas are highly valued by society. Water management or mismanagement has often contributed to the ecological degradation of wetland-riparian areas. In the past, such sites were often abused by excessive grazing and/or promiscuous burning. Weed infestations in these areas were often ignored. Past abuse of such habitats paved the way for rapid expansion of invasive weed species.

L. latifolium is extremely tolerant to halomorphic soils. Both in coastal areas and interior salt marshes and sinks, this weed has proven to be highly adaptable. *L. latifolium* is a major problem in brackish marshes of the San Francisco Bay in California and in saline/alkaline sinks in the Carson Desert of western Nevada. Usually these would be considered very contrasting macro-environments, but the ability to compete on extreme halomorphic soils allows the weed to invade, colonize and dominate these contrasting environments.

Invasion mechanism

In the Intermountain Area of the western United States, initial establishment of *L. latifolium* has often occurred in association with structures for diversion of irrigation water from streams. Construction or repair of such physical facilities usually provides bare soil for colonization by weeds. The undercarriage of track-laying heavy construction equipment is an ideal mechanism for transporting vegetative propagules of weeds from site to site. Once *L. latifolium* becomes established in a riparian system it can spread laterally by both seed and creeping rootstocks. The rootstocks are buoyant and when broken from eroding banks can be transported long distances by water.

L. latifolium has the potential for tremendous seed production once fully established. The role domestic livestock and waterfowl play in dispersing these seeds is not fully understood. There is much concern that *L. latifolium* seeds are being transported in high quality alfalfa (*Medicago sativa* L.) grown in the Intermountain Area and transported to California dairy farms. In well-managed alfalfa hay fields, *L. latifolium* does not appear to be a contaminant and the cutting and harvesting practices for alfalfa usually preclude seed contamination of hay. Alfalfa hay is often stacked in yards before being transported long distances and these yards are often infested with *L. latifolium*. Long distance seed transfer may be from truck tires and undercarriages rather than from direct contamination of hay. Long distance transportation of straw from grain crops may be a source of *L. latifolium* spread. Straw is often used to reduce erosion on construction projects. This well-intentioned use of straw to mediate environmental disturbance may often serve to inoculate disturbed sites with noxious weeds. In the case of *L. latifolium* the straw may be weed free, but the storage sites for the material may provide the inoculum with the transportation vehicles serving as the vector for the weed.

In Intermountain area agriculture, the development of *L. latifolium* infestations, after initial establishment in riparian areas, has often consisted of infestations moving along irrigation ditches and canals to agricultural fields. A major type of forage production in the region is native hay production from irrigated meadows. This is accomplished with natural flooding of extensive areas on the floodplain along streams and then usually making only one cut for use as winter forage in range

Table 1. Germination profile for four replications of seeds of *L. latifolium* at 55 constant and alternating simulated day and night temperatures (Miller *et al.* 1986).

	Germination Warm-period temperature (°C)									
Cold-period temperature	0	2	5	10	15	20	25	30	35	40
(°C)	(%)									
0	0	0	0	1	5	15	47	75	84	77
2		1	5	28	83	95	92	96	91	85
5			15	21	29	58	98	92	93	86
10				5	15	14	89	99	94	98
15					6	18	39	80	97	99
20						9	12	56	95	94
25							20	18	53	91
30								21	28	39
35									23	20
40										7

livestock production. The meadows are a mixture of native sedges (*Carex* spp.), grasses, and a suite of adventitious grasses. Irrigation structures in these meadows are crude earthen embankments that must be frequently renewed. This annual disturbance provides ideal habitat for the establishment of *L. latifolium*. Once established, this weed can successfully compete with the native meadow species and within a decade the result is nearly total dominance of entire fields. The economic consequences of this invasion are reduced forage quality in terms of digestible nutrients, especially available protein (Young *et al.* 1995). The native hay meadows serve as nesting habitat for waterfowl during the spring period when they are flooded. Invasion by *L. latifolium* changes the cover and food production characteristics of these meadows for the birds. Waterfowl habitat managers are among those most concerned by the consequences of invasion by this weed, but no one has measured and quantified the impact of this weed on waterfowl habitat.

In the Susan River Valley of northeastern California, *L. latifolium* invasion has proceeded beyond the native meadow to where the weed is now a serious pest in more intensive agricultural rotation crops such as cereal grains and alfalfa. The difficulty of controlling *L. latifolium* in the alfalfa portion of this rotation, which is the highest cash value crop, has led to depreciation in land values for infested farms (personal communication from Jay Dow, Tule Farms, Wendel, California).

Fig. 1. Accumulation of semi-woody biomass in *L. latifolium* stands.

Life cycle

Seedlings of *L. latifolium* are difficult to recognize in the field. The pale green cotyledons are easily confused with many other adventitious members of the *Brassicaceae* family. Laboratory tests have shown that seeds can be highly viable and germinate over a wide range of constant and alternating temperatures (Miller *et al.* 1986) (Table 1). Field studies in Lassen County, California, have shown the annual seed rain to be over 16 billion seeds per hectare (unpublished data, USDA, ARS, Reno, NV). In dry years, flowering of *L. latifolium* is profuse, but seed set and maturity is minimal. In very wet years, rampant infection with the white rust (*Albugo* sp.) appears to largely inhibit seed production. Attempts to determine the nature of *L. latifolium* seedbanks have proven inconclusive. In well established stands, the seeds apparently have a very short half-life.

Rather than expansion from seedlings, it is much more common to see *L. latifolium* populations expand by creeping rootstocks advancing 1 to 3 m from the main colony and sending up shoots (unpublished research, USDA, ARS, Reno, NV). In the western Intermountain Area, shoot growth is apparent in late March. Even during mid-winter, careful examination of root crowns reveals multiple buds that are green and very slowly developing. Initially, the shoots form a rosette near the soil surface. These rosettes remain largely hidden because of the persistent, semi-woody herbage on the stands from previous years' growth. These stems accumulate over several years, forming nearly impenetrable thickets. This characteristic of persistent

Fig. 2. L. latifolium flowers.

semi-woody stems negatively influences many potential control measures for the species (Fig. 1).

Stem elongation is rapid during May and by the first of June stems are 0.5 to 0.8 m in height. Inflorescence with budding is apparent by mid-June. At flowering, the stalks are 1 to 2 m in height. The long petiolate, wide-bladed early leaves that formed a basal rosette are nearly senescent by flowering time. The cauline leaves are much reduced, with short petioles on the lower stem and sessile toward the inflorescence. The reduction in leaf surface by flowering may play a role in the difficulty in controlling this weed with foliar active herbicides typically applied at flowering. At flowering time, the stands are usually so thick that virtually no light reaches the soil surface.

Typical dense colonies present a solid array of very small, typical *Brassicaceae* family, white flowers (Fig. 2). The flowers are arranged in clusters on the panicle branches (Fig. 3). The seeds are mature by mid-August, but plants mowed in hay fields or injured in weed control treatments will flower late and there may be seeds

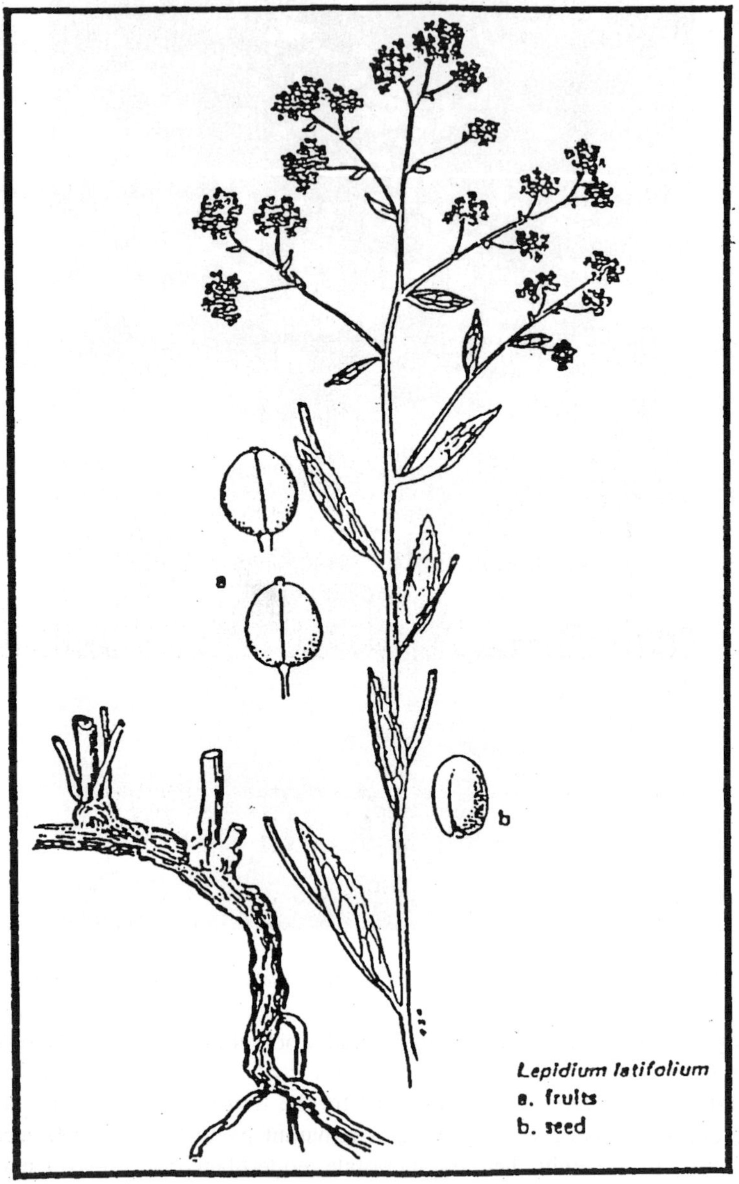

Perennial Pepperweed

from Gilkey's Weeds of The Pacific Northwest

Fig. 3. L. latifolium panicle and flower cluster structure.

Fig. 4. L. latifolium seeds on a 1 mm grid.

still maturing by the first frost. The mature seeds do not immediately disperse from the pods. Some seeds remain on the plants until the next season (Fig. 4).

Consequences of *Lepidium latifolium* invasion

It is commonplace to refer to dense infestations of weeds as near monocultures. In the case of well established stands of *L. latifolium,* this is no exaggeration (Fig. 5). Apparently this noxious weed accomplishes dominance the old fashioned way; through competition for moisture, nutrients and light. There is no verifiable evidence of allelopathy (Blank and Young, unpublished data). *L. latifolium* does influence the nature of surface soils through a tremendous build up of organic matter from annual biomass production. The stems, as previously mentioned, are semi-woody and take many years to decay. The leaf fall and finer stems build a dense organic layer on the soil surface (Blank and Young 1997). The consequences of this build up on carbon-nitrogen ratios over time may be very significant. This weed is adapted to using water containing high levels of salts. This results in a build up of these salts in the plant biomass. Essentially *L. latifolium* growing on halomorphic soils acts as a salt pump, removing salts from the soil solution and depositing them on the soil surface. This may be extremely significant over prolonged periods.

The ultimate significance of *L. latifolium* colonization is the effect on subsequent seedling recruitment and succession. In wetland-riparian areas, woody species of *Populus* and *Salix* often provide critical habitat for dependent animal species. We do

Fig. 5. Dense stand of *L. latifolium* along the east Truckee River in Reno, Nevada.

not understand seedling recruitment in these communities without *L. latifolium* dominance. When this noxious colonizer is present, seedling recruitment becomes even more complicated.

Control methods

It is quite obvious that in western North America *L. latifolium* is so widely distributed and so abundant that eradication is economically and perhaps biologically impossible. The goal must be to apply integrated pest management procedures to biologically suppress *L. latifolium*.

Young *et al.* (1995) pointed out that there were no successful examples in the world of classical biological control of a member of the family *Brassicaceae*. This has been criticized as too sweeping a statement. Perhaps the statement is true because no one has ever tried using biological control on the *Brassicaceae* family (personal communication from P. C. Quimby, Biological Control of Weeds, USDA, ARS).

Mechanical control of *L. latifolium* interacts unfavorably with the abundant creeping rootstocks of this species. It is virtually impossible to cut the roots in small enough segments so that a significant number will not produce new plants from latent buds (Wotring *et al.* 1995) (Table 2).

Suppression of weeds through livestock can be a very useful tool in integrated management. When *L. latifolium* plants occur as occasional plants in saltgrass (*Distichlis spicata*) meadows, we have noted significant use of flower stalks by

Table 2. Shoot growth characteristics of *L. latifolium* from five replications of 2.5 cm-long root sections with various diameter size ranges (Wotring *et al.* 1995).

Root diameter range (cm)	Avg number of days until first sprout[1]	Avg number of sprouts per root[1]	Avg sprout length (cm)[1]
0.05–0.75	20	0.90	4.8
1.00–1.25	15	0.40	10.0
1.50–1.75	19	0.90	5.7
2.00–2.75	16	0.90	19.1
3.00–4.00	18	0.70	14.2

[1]There are no significant differences among root diameter range sizes for any of the three dependent variables measured, based on nonsignificant F-values from three one way analyses of variance.

grazing cattle. In dense stands of the weed in irrigated pastures, there is no evidence of utilization by grazing animals even when drought had seriously depleted preferred forage species. Reportedly, cattle will graze the rosettes of leaves in the early spring. Unless the accumulation of previous years' growth of stalks is mechanically removed in the fall, these leaves cannot be reached by livestock. There have been persistent reports that *L. latifolium* forage is poisonous to livestock. Feeding trials are underway at the Poisonous Plant Research Laboratory, USDA, Logan, Utah, to ascertain if and under what conditions consumption of this plant could be harmful to livestock.

The aerial portion of *L. latifolium* plants are susceptible to applications of several herbicides. Generally, the plants are most susceptible to foliar active herbicides at the initiation of flowering. The problem with these herbicidal treatments has been rapid renewal of the stands from the creeping rootstocks. We have determined that even with 98% treatment control, resprouting plants in the spring will result in total stand dominance by the end of the growing season.

The herbicide chlorsulfuron, 2-chloro-N-[[(4-methoxy-6-methyl-1,3,5-triazin-2-yl)amino]carbonyl]benzenesulfonamide, which is both foliar and soil active, is effective in controlling *L. latifolium* when applied at the correct rate and time (unpublished research USDA, ARS, Reno, NV). This herbicide does not kill grass and grass-like species at the rates needed for control of *L. latifolium*. But, chlorsulfuron effectively kills the valuable woody species that occur in wetland-riparian areas where restoration is often the goal of *L. latifolium* control programs. We do not have enough experience with this herbicidal technique to know how long and under what conditions it will remain effective and how tolerant both crop and native species will be to soil active residues.

Any lasting biological suppression of *L. latifolium* is dependent on recruiting seedlings that are capable of competing successfully with the weed in managed ecosystems. In wetland-riparian systems perhaps we should be looking toward competitive ecotypes of native species to accomplish this goal. An example would be the use of saltgrass in halomorphic wetland areas.

References

Blank, R.R. and J.A. Young. 1997. *Lepidium latifolium*: Influences on soil properties, rate of spread, and competitive stature. This volume.

Miller, G. K., J. A. Young, and R. A. Evans. 1986. Germination of seeds of perennial pepperweed (*Lepidium latifolium*) Weed Sci. 34:252-255.

Robbins, W.W. 1940. Alien plants growing without cultivation in California. Calif. Agr. Expt. Sta. Bull. 637. 128 pp.

Young, J.A., C.E. Turner, and L.F. James. 1995. Perennial pepperweed. Rangelands 17:121-123.

Wotring, S.O., D.E. Palmquist, and J.A. Young. 1995. Shoot growth from creeping rootstocks of perennial pepperweed. Abst. Society for Range Management Annual Mtg., Jan. 16-19, 1995, Phoenix, AZ.

LEPIDIUM LATIFOLIUM: INFLUENCES ON SOIL PROPERTIES, RATE OF SPREAD, AND COMPETITIVE STATURE

Robert R. Blank and James A. Young
USDA-Agricultural Research Service, Conservation Biology of Rangelands Unit, 920 Valley Road, Reno, NV 89512, USA

Abstract

Lepidium latifolium (perennial pepperweed, tall whitetop) is an Eurasian crucifer that is aggressively invading wetlands and riparian habitats across the western United States. During the past four years we have investigated the influence of *L. latifolium* on the physical and chemical properties of soil and the nature of its competitiveness using field and greenhouse studies. The rapid spread of this invasive plant is governed by seed dispersal through water and creeping rhizomes. In seasonally flooded soils along the Susan River of northeastern California, the invasion of *L. latifolium* has caused firm and compact natric horizons to become friable and less distinct. Greenhouse studies demonstrated that as soil fertility levels decline, the growth potential of *L. latifolium* significantly ($P \leq 0.05$) declines and that it allocates more resources to roots. When competing against *Bromus tectorum*, an aggressive annual grass, there is a threshold of soil fertility whence *L. latifolium* cannot obtain soil nutrients as efficiently as *B. tectorum*, at which time, *B. tectorum* has superior growth potential. In established monocultures of *L. latifolium*, shading and accumulation of plant litter may contribute to its competitiveness. Monocultural stands of *L. latifolium* may be self-limiting as soil fertility levels decline. To better predict invasiveness and longevity, research is needed to quantify nutrient cycling, particularly nitrogen, in *L. latifolium* dominated ecosystems.

Introduction

Lepidium latifolium L. (perennial pepperweed, tall whitetop) is a crucifer native to southeastern Europe and Asia that has become widely distributed in western North America and in coastal New England. During the past two decades, the species has spread rapidly in the Pacific coastal and intermountain states. A survey conducted during the summer of 1993 revealed that *L. latifolium* is a major weed in wetlands in all western states except Arizona (Young *et al.* 1995). Wetlands and riparian habitats are now recognized as being extremely important components of ecosystems. The weed is found in all but three counties in California, and in a contrasting environment, all but two counties in Nevada. It is now obvious that *L. latifolium* has the potential to spread to native hay meadows that are an essential part of livestock production systems in much of the intermountain and mountainous areas of the western United States. *L. latifolium* is also a serious weed concern in northern Europe (Lyeik 1989) occupying virtually every landfill site (Kiegel *et al.* 1988). This weed is also a problem in Australia (Kloot 1973).

L. latifolium plants are from 0.4 to 1.0 m tall, with a multitude of stems forming

Plant Invasions: Studies from North America and Europe, pp. 69–80
edited by J.H. Brock, M. Wade, P. Pyšek and D. Green

dense thickets. The plants have extensive underground rhizomatous systems that radiate in all directions from a newly established plant. Aerial shoots emerge from buds along these spreading rhizomes. In two seasons, a single established plant becomes a small colony, up to several meters in diameter. Tillage results in multiple sprouts from cut rhizomes. In the Honey Lake Valley of northeastern California, seed production from stands of 200 stems m^{-2} was calculated to be 16 billion seeds ha^{-1} (Young *et al.* 1996). Field infestations of *L. latifolium* form almost pure colonies with virtually no species diversity. Its height and density suppresses light reaching the soil surface, which may be a factor in its competitive potential. Although the weed dies back annually, old stems remain to shade the soil surface. In habitats where *L. latifolium* invades, it generally replaces perennial grasses including highly competitive species such as *Elytrigia repens* (L.) Nevski.

The goals of this paper are to: 1) describe the rate of spread of *L. latifolium* and examine the plant strategies that enable this invasive species to form monocultures; 2) determine if the invasion of *L. latifolium* differentially changes soil characteristics as compared to native plant communities, and if so, what are the potential long-term consequences of such changes; 3) quantify changes in soil fertility levels caused by *L. latifolium* invasion and determine if lowered fertility levels may be self-limiting to this species; and 4) gauge the competitive potential of *L. latifolium* against the highly competitive annual grass *Bromus tectorum*.

Materials and methods

Rate of colonization study

Research was conducted at the Honey Lake Wildlife Refuge, California, approximately 74 km north-northwest of Reno, Nevada. Honey Lake is a remnant of an immense Pleistocene lake which covered much of northwestern Nevada and adjacent California. During high water periods, lacustrine sediments accumulated. Parent materials for most of the soils around Honey Lake are fine-textured sediments along with aeolian deposits and reworked, coarse-textured, beach and offshore deposits. Our study area had been planted to *Elytrigia elongata* (Host) Nevski eight years earlier as part of the management plan for the wildlife refuge. This area is rarely flooded naturally, but is often flood-irrigated. A permanent 40 m by 40 m plot was established in 1994. At that time, there were only four small colonies (< 1 m diameter) of *L. latifolium*. Using a 1 m^2 frame (×4), we estimated the number of *L. latifolium* stems per 4 m^2 in a grid pattern. Stems were counted in June of 1994, 1995, and 1996.

Plant-soil relationship study

This particular study was conducted approximately 2 km north of the Honey Lake Wildlife Refuge. Lacustrine sediments are the dominant parent material for the soils, but a high sodium content has caused strongly-expressed, firm and compact subsoils or natric horizons to form. The topography is nearly level to very gently undulating, and interrupted by low swales associated with the distributary system of the Susan River. The area is frequently flooded from spring snowmelt off the Sierra-Cascade mountains to the west via the Susan River. We compared pedons (two replicates) in an area that had been invaded by *L. latifolium* for over 5 years to an adjacent uninvaded area of native hayland occupied by *Juncus* sp., *Pascopyrum smithii* (Rydb.) A. Löve, and *Distichlis spicata* (L.) E. Greene. The soils were field-described and sampled using established protocols (Soil Survey Staff 1984). Soil brought to the laboratory for study was air-dried and lightly crushed to pass a 2 mm sieve. The soil was homogenized and the following attributes measured: organic C by dichromate digestion (Nelson and Sommers 1982); Kjeldahl-N by catalyzed digestion in sulfuric acid (Bremner and Mulvaney 1982) with quantification of ammonium by membrane diffusion flow injection; and electrical conductivity (EC) from a saturated paste extract (Richards 1954).

Root length per volume of soil was determined from cores taken at 0–10 and 10–20 cm depths in *L. latifolium* invaded and non-invaded areas (3 replicates). The soil was dispersed in sodium hexametaphosphate and then repeatedly washed with a commercial root washer until the roots were completely separated from soil material. The total length of root was quantified with a commercial root scanner. No corrections were made for the amount of dead roots. The length of root was divided by the volume of soil to obtain root length in m cm^{-3} of soil.

Competition and soil nutrient study

The purpose of this greenhouse study was to investigate the competitiveness of *L. latifolium* and growth changes related to soil fertility levels. We evaluated two soil types: a low fertility coarse-textured soil and a high fertility fine-textured soil obtained on the University of Nevada, Reno, Experiment Station that is presently being invaded by the weed. The treatments applied to the two soils were: a) *L. latifolium* (LELA); b) *L. latifolium* + *Bromus tectorum* (LELA + BRTE); c) *B. tectorum* (BRTE); and d) control (unplanted). *B. tectorum* is an invasive alien annual grass that has become dominant in many uplands throughout the sagebrush-steppe of the western United States. The pots contained approximately 15 kg of low fertility soil and 8 kg of high fertility soil (equivalent volume of soil). *L. latifolium* was planted as a 2.5 cm long rhizome section, one section per pot. The sections were freshly collected from an area where the fertile soil was collected. Three seeds of *B. tectorum* were planted in each pot. In the combined treatment, *L. latifolium* was planted in the center of the pot with three *B. tectorum* seeds equally spaced around. Pots were kept continuously moist with deionized water. A plastic plate was placed beneath the pots to prevent loss of water or elements.

During the course of the experiment, plant height and total number of leaves were recorded intermittently. In one of the treatment replicates, a ceramic tension lysimeter was installed and soil solution was extracted periodically throughout the period of plant growth. Ion chromatography was used to quantify solutes in tension lysimeter extracts. The experiment was concluded after *L. latifolium* flowered. Above ground mass was clipped at the soil surface, dried at 60°C and weighed on an individual plant basis. Roots were carefully extracted from the soil in the pots with observations and photographs taken of the location of roots in pots. The roots were washed until free of soil, dried at 60°C and weighed. Because of the large size of *L. latifolium* roots, we obtained near complete recovery of roots, however, many very fine roots of *B. tectorum* could not be separated from the soil. Pot soil was air-dried

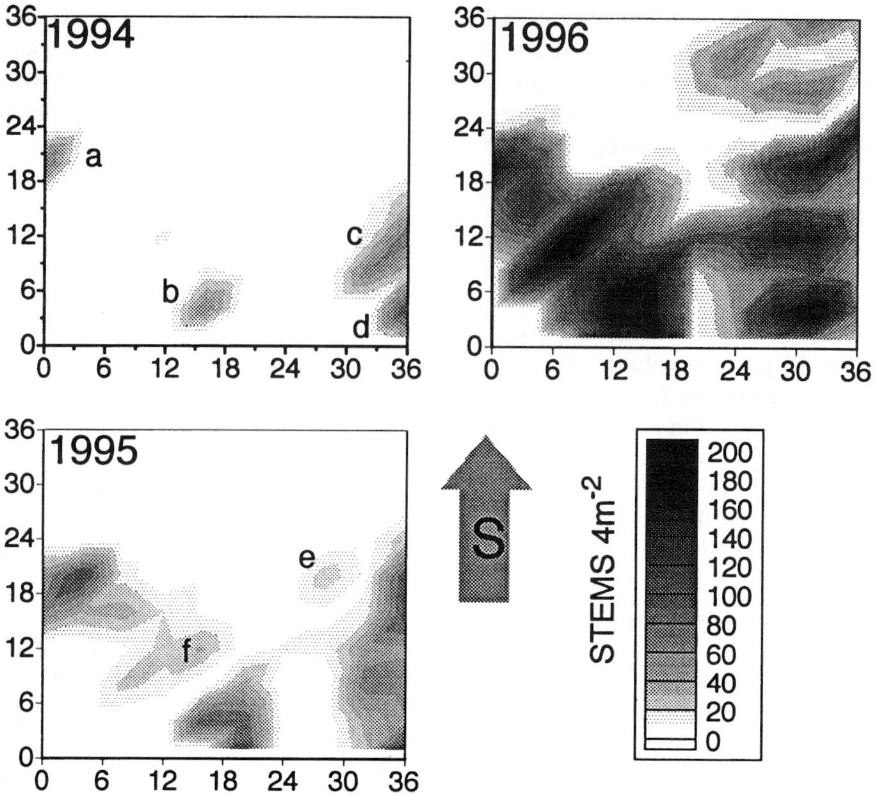

Fig. 1. Contour fill graph showing density of *L. latifolium* stems per 4 m^2 in 1994, 1995, and 1996. Units of the x and y axes are in meters.

in the greenhouse, homogenized, and a portion passed through a 2 mm sieve and stored for later laboratory analyses. The remaining soil was combined and homogenized, by treatment, then re-potted and a new experiment was begun re-applying the same treatments, for a total of 3 growth cycles. The first growth cycles had 4 replicates; for the second and third growth cycles there were 3 replicates. To keep photoperiod constant during the three experiments, artificial lighting was used to increase daylight hours during all of the first growth cycle (Jan. 4 to Mar. 8, 1995) and a portion of the second growth cycle experiment (Mar. 14 to May 22, 1995). No artificial lighting was used during the third growth cycle (June 1 to Sept. 8, 1995). During the experiment, the greenhouse temperature was set at 90°C maximum and 50°C minimum. During the first growth cycle the maximum temperature was rarely achieved.

Dried leaves and roots from each replicate of the experiment were separately ground in an Udy-mill and analyzed for total N. For some samples, particularly for *B. tectorum* grown in non-fertile soil, there was insufficient plant material. On air-dried, < 2 mm homogenized material of the original soils, and all treatments and growth cycles of the pot soils, the following attributes were measured: KCl-extractable nitrate and ammonium (Keeney and Nelson 1982) and bicarbonate extractable ortho-P (Olsen and Sommers 1982).

Lysimeter data were analyzed using regression techniques. Treatment attributes were compared at each time using confidence intervals about the attribute estimates. Plant and soil attributes data were analyzed using a mixed model 3-way ANOVA. If the F-test was significant, means were separated using Duncan's new multiple range test.

Results and discussion

Rate of colonization study

The rapid rate of spread of *L. latifolium* is evident from Figure 1. In 1994 there were four individual colonies marked a, b, c, d. By 1995 all colonies had expanded, colonies c and d had merged, and there were two new colonies (e and f). Vividly apparent in 1995 was the high recruitment of plants between colonies a and b. Close examination of this area revealed a debris trail, and scouring and deposition patterns indicative of flooding. In 1996, the densest colonies of *L. latifolium* contained over 50 stems per m^2, nearly double that of 1995. Monocultural stands of *L. latifolium* occurred when stem density exceeded about 20 per m^2. Excavation of plants in these colonies showed that most stems emanated from rhizomes. Judging from examination of invasion fronts occurring throughout the Honey Lake Wildlife Refuge, the invasion process has two modes. If the invasion is occurring over undisturbed native vegetation, the front is narrow with the dominant mode of invasion via creeping rhizomes. In more disturbed areas, the invasion front is very wide with small (< 1 m diameter) colonies scattered throughout. A visual estimate of *L. latifolium* coverage at this initial stage of invasion would be approximately 2 percent. A conceptual model to describe the invasiveness of *L. latifolium* involves initial long-scale migration via water-borne transport of the seed pods. Seeds with pods removed are highly viable and germinate at a wide range of constant and alternating temperatures and no light requirement (Miller *et al.* 1986). These seed-borne populations then spread by creeping rhizomes and establish single species colonies. Finally, colonies expand, eventually merging. A key to its competitiveness in colony form is the large amounts of reserves stored in perennating organs that enable the plant to rapidly produce foliage and thus a high density of plant cover (Grime 1974). This results in reduced sunlight at the soil surface providing an additional competitive advantage (Radosevich and Holt 1984).

Plant-soil relationship study

There were several notable differences in soil properties between *L. latifolium* invaded and non-invaded control areas (Table 1). *L. latifolium* invaded areas have thick, relatively undecomposed litter layers that are absent in control areas. These litter layers can have extraordinarily high levels of water-soluble ammonium exceeding 100 mg kg^{-1} of soil (unpublished data 1993). *L. latifolium* litter has a very high N content and low C:N ratios (Table 1). The presence of *L. latifolium* appears to have altered the strength and structure of the natric horizons. Non-invaded native haylands exhibit firm and compact sodium-affected subsoils (Btn horizons). These subsoils in *L. latifolium* invaded areas consist of a very black friable layer with visible crystals

Table 1. Soil descriptions and selected attributes for *Lepidium latifolium* invaded and non-invaded control areas[1].

Horizon	Depth	Texture	Color		Structure	pH	EC	% N	%C	C:N
Invaded area										
O	0–+5					7.37	1.3	2.45	32.7	13.3
A	0–15	sil	10YR	3/2	granular	7.78	5.8	1.55	14.2	9.2
Bty	15–28	sicl	7.5Y	2/0	prismatic	7.63	8.0	0.33	3.8	11.1
Bt	28–56	sicl	10YR	3/2	prismatic	7.49	12.0	0.03	0.4	15.1
BC	56–76	sicl	10YR	4/2	blocky	7.37	11.8	0.02	0.3	12.8
C	76–107+cl		7.5YR	4/2	blocky	7.04	22.0	0.03	0.3	10.4
Non-invaded area										
A	0–6	sicl	2.5Y	2/0	granular	8.01	5.4	0.99	14.0	14.2
Btn	6–18	sic	2.5Y	2/0	columnar	7.85	3.4	0.27	3.5	12.9
Bt1	18–33	sicl	2.5Y	5/2	prismatic	7.83	2.5	0.03	0.5	18.8
Bt2	33–48	sicl	2.5Y	5/2	prismatic	7.95	2.4	0.01	0.4	11.0
BC	48–69	sicl	2.5Y4.	5/2	blocky	7.90	3.6	0.01	0.4	59.7

[1] There were no discernable differences in landscape position between the invaded and non-invaded areas: thus, we submit that the soils were similar before the invasion. To support the initial finding, the pedons described in the non-invaded area will be described and sampled again after this site has been invaded by *L. latifolium* for 5 years. In 1995, the non-invaded area was nearly completely occupied by *L. latifolium*.

of salt. The root length density of *L. latifolium* is less than the root length density of the native community (Fig. 2). No attempt was made to separate roots by species in this study. Moreover, the measured values in the *L. latifolium* invaded area assuredly contained roots from native species which over-estimated the *L. latifolium* roots. The rooting system of *L. latifolium* is designed for rapid uptake of nutrients obtained via mass flow; it is a poorly adapted system for exploration of nutrients.

Soil, the climate to which it is exposed, and the flora and fauna it supports are inexorably linked (Jenny 1941). Van Breemen (1993) argued that biotic processes make the soil more favorable for plant growth which is analogous to the Gaia hypothesis of Lovelock (1979). Can the invasion of an area by a species such as *L. latifolium* alter soil properties to such an extent that different successional trajectories are triggered, and if successional trajectories are changed, will soil evolution correspondingly diverge? El-Ghareeb (1991) reported that, in Egypt, the alien plant *Mesembryanthemum crystallinum* so increases the osmotic potential of the seedbed that native annuals are unable to germinate. The invasion of the western United States by *Bromus tectorum* has the potential to decouple soil-plant processes in which soil development occurs under shrub-bunchgrasses, processes that have operated since the Pliocene. By increasing competitive pressure and fostering more frequent wildfires, this annual grass has converted diverse biological communities into monocultures (Billings 1992).

What are the potential effects of *L. latifolium* invasion on soil properties? It is simply too early to predict. One might expect that the amelioration of dense compact subsoils could have a positive effect on plant growth. Moreover, decomposition of the nitrogen rich litter would seem favorable in building thick, more organically rich surface soil horizons. However, it is unclear if the rooting pattern of this plant will

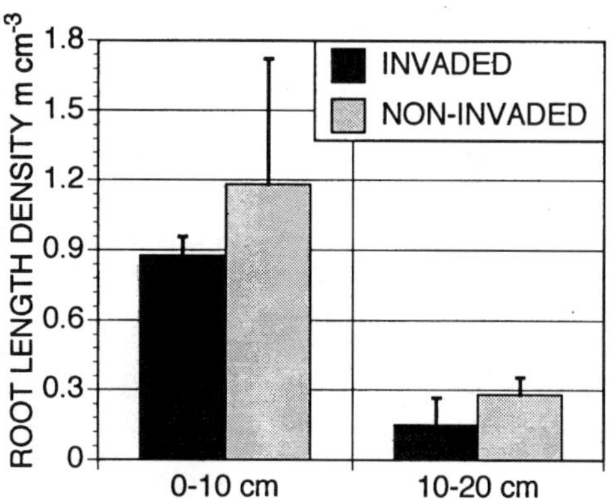

Fig. 2. Root length in soil collected from a *L. latifolium* invaded area and adjacent native hayland. Standard errors are displayed for this and following figures.

result in greater salt accumulation in the soil surface. It is also unclear how nutrient cycling, particularly nitrogen cycling, will differ in *L. latifolium* monocultures compared to diverse native communities.

Competition and soil nutrient study

When grown by itself, or in competition with *B. tectorum*, *L. latifolium* significantly ($P \leq 0.05$) declined in above ground mass from the first to the second and third growth cycles (Fig. 3a). When grown in the fertile soil, except for the first growth cycle, *B. tectorum* produced as much above ground mass as *L. latifolium*. In the non-fertile soil, however, growth of *L. latifolium* outpaced growth of *B. tectorum*. For the second and third growth cycles in the fertile soil, competition with *B. tectorum* significantly ($P \leq 0.05$) reduced the above ground mass of *L. latifolium*. As one would expect, growth of *L. latifolium* and *B. tectorum* was greatest in the fertile soil.

When grown in the fertile soil, shoot/root ratios of *L. latifolium* declined significantly ($P \leq 0.05$) and precipitously from the first to the second and third growth cycles (Fig. 3b). For *B. tectorum*, shoot/root ratios were variable with growth cycle. Given the loss of roots in the process of separating *B. tectorum* roots from soil, the ratios are likely somewhat lower. During the second growth cycle for the fertile soil, competition with *B. tectorum* reduced the shoot/root ratio for *L. latifolium* as compared to the *L. latifolium* on its own. In the first growth cycle of the non-fertile soil, *L. latifolium* showed extremely high shoot/root ratios (Fig. 3b). In the second and third growth cycles, however, the shoot to root ratios declined significantly ($P \leq 0.05$) to values near one.

The distribution of *L. latifolium* roots in the pots is illustrative of its mode of resource acquisition. The longest roots averaged over 1.2 m in length and most of that length was curled around the bottom of pots. The amount of soil explored by *L. latifolium* was minimal which suggests that the principle pathway of nutrient transport to its roots is via mass flow (Barber 1995).

In most treatments, there was a significant decline in tissue N content with growth cycle (Fig. 3c). Given the decline in average soil inorganic N content during the same period (see Fig. 4), these results are appropriate. One surprising result is that

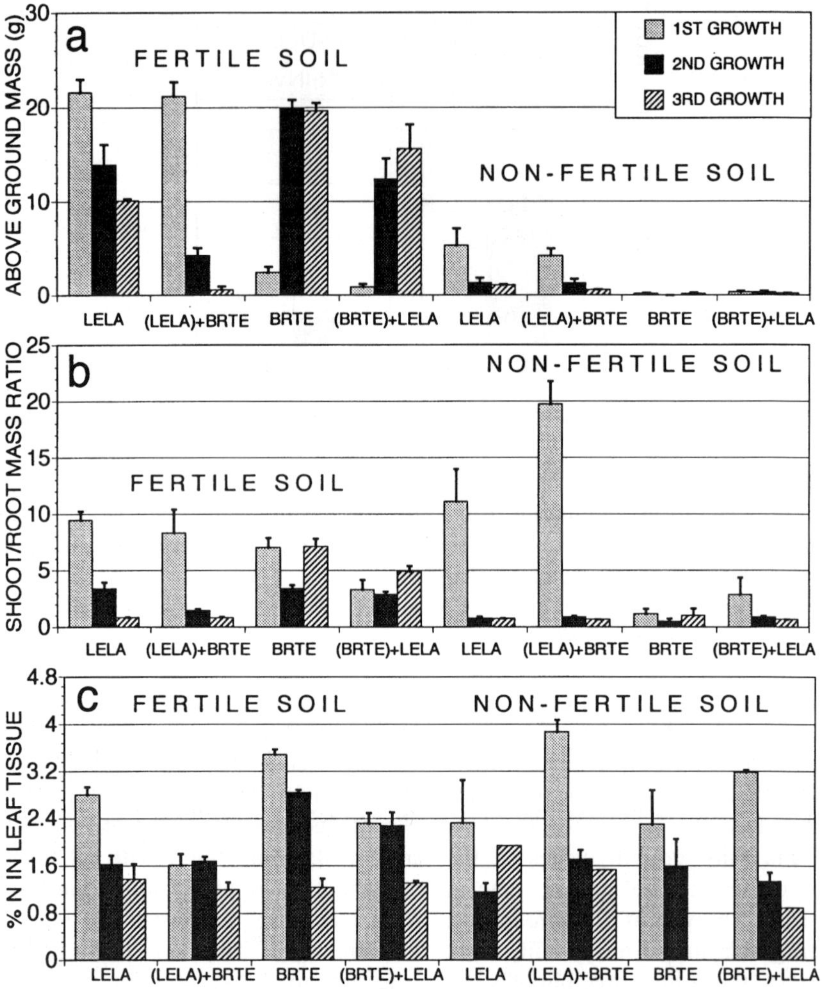

Fig. 3. Graphs showing significant (P ≤ 0.05) treatment × soil fertility level × growth cycle interaction in aboveground masses (a), root to shoot mass ratios (b), and percent N in leaf tissue (c) of *L. latifolium* and *B. tectorum*. *L. latifolium* shown in (LELA)+BRTE treatment; *B. tectorum* shown in (BRTE)+LELA treatment.

the plants grown in non-fertile soil had a similar N content as did the plants grown in fertile soil. This finding is perplexing. If the fertile soil declined in N, then why should the non-fertile soil have high N content when it has extremely low levels of soil N. We can only conclude that the lower production of biomass compensated for this.

For the fertile soil, levels of nitrate in lysimeter extracts are remarkably high and can partially explain why *L. latifolium* is successfully invading these soils (Fig. 4). As one would expect, in most of the planted treatments, levels of nitrate decline during plant growth. In the first growth cycle, however, the decline was similar to that of the control. This finding suggests that denitrification may be taking place in the control pot (and other treatments) which is plausible given the high water appli-

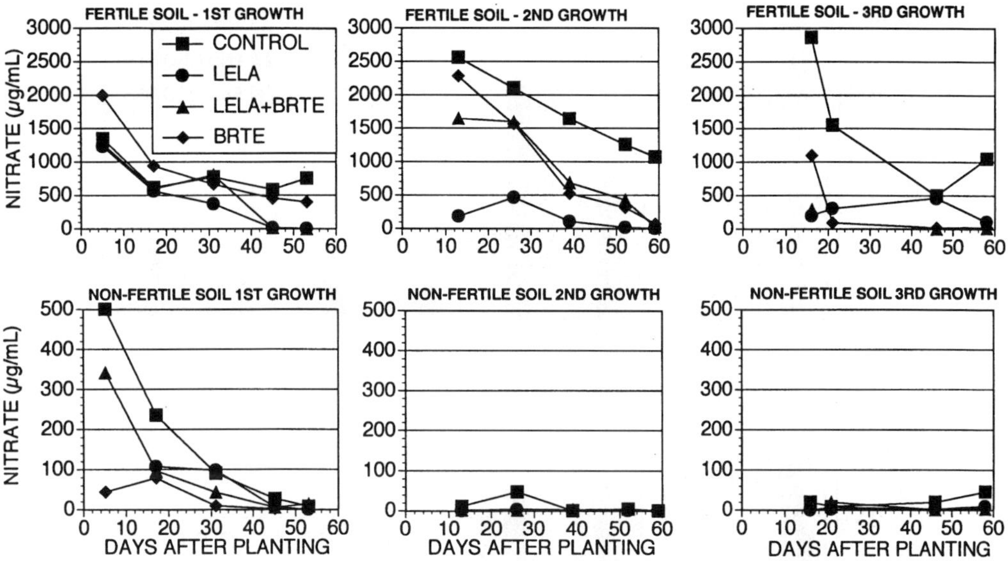

Fig. 4. Concentration of nitrate from lysimeter extracts taken at various times after planting as influenced by treatment, soil fertility levels, and growth cycle.

cation rates which are likely to create anaerobic microsites. In the second and third growth cycles, there was more differentiation between the control treatment and other treatments with the *L. latifolium* treatment lowering nitrate content the most. Another interesting aspect is that nitrate for the fertile soil increased from the end of one growth cycle to the beginning of the next. We suspect that this is due to soil drying in this high organic matter soil which fosters rapid N mineralization (Birch 1958).

Levels of nitrate in lysimeter extracts for the non-fertile soil declined rapidly in the first growth cycle, then remained low in the second and third growth cycles. The initial nitrate levels in the non-fertile soil were very high and may have been due to long-term stockpiling outdoors near the greenhouse facility. After initial plant uptake of latent nitrate, subsequent low levels suggest very low mineralization potentials. Similar to the fertile soil, levels of nitrate in the non-fertile control treatment declined rapidly which suggests that denitrification was occurring.

As one would expect, the fertile soil had far higher levels of extractable nitrate than did the non-fertile soil, with the control soil reaching nearly 1000 mg kg^{-1} (Fig. 5a). For the planted treatments, extractable nitrate generally declined with growth cycles, with greatest decline being that for the non-fertile soil. For the fertile soil in the first and second growth cycles, *L. latifolium* on its own, and *L. latifolium* and *B. tectorum* combined treatments significantly reduced extractable nitrate as compared with the *B. tectorum* treatment. Given the amount of plant growth in the fertile soils in the second and third growth cycle, the levels of extractable nitrate seemed extraordinarily high. Clearly the rate of nitrogen mineralization is very high in this soil to have nearly 100 mg kg^{-1} of extractable nitrate after three successive growth cycles. The non-fertile soil had far lower levels of extractable ortho-P than the fertile soil

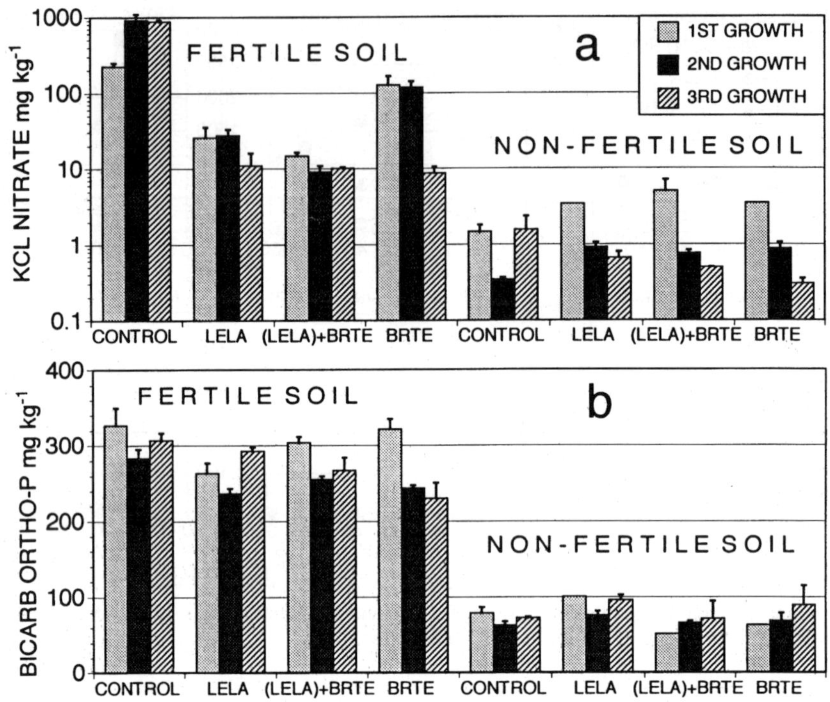

Fig. 5. KCl-extractable nitrate (a) and ortho-P (b) from pot soils as influenced by significant ($P \leq 0.05$) treatment × soil fertility level × growth cycle interaction. Data for nitrate is presented on a logarithmic scale to show differences in the non-fertile soil.

(Fig. 5b). In the fertile soil, levels of extractable ortho-P declined relative to the control soil for all treatments as growth cycle increased. The significant treatment interaction was due to slightly lower extractable P levels in soil planted to *B. tectorum* as compared with those planted with *L. latifolium.* This finding is contradicted in the non-fertile soil where levels of extractable ortho-P increased with growth cycle in treatments where *B. tectorum* is grown.

General discussion and conclusions

L. latifolium would fit Grime's (1977) category as an ideal weed. It is both a prolific seed producer and can rapidly expand by creeping rhizomes. Moreover, its ability to spread and form dense monocultural stands which retard light penetration to the soil surface would be another competitive advantage (Radosevich and Holt 1984). In established stands, the thick litter-rich layers of relatively undecomposed plant material are likely to hinder the germination of seeds of other species and further enhance its competitive potential. Our data, however, hints at a natural limitation to its competitiveness.

Roots of *L. latifolium* are highly elongated and thick, with minimal branching. In pots, the volume of soil explored is small. Given these root characteristics, it appears that mass flow of nutrients to the root is a controlling factor in the mineral nutrition of *L. latifolium* (Marschner 1986). Resource acquision via mass flow may explain why *L. latifolium* is most invasive in wetland soils; the soil is generally moist enough to allow efficient mass flow of nutrients to the root without the encumbrances of tortuosity or nutrient impedance as would come into play in drier soils (Barber 1995). When grown individually the above ground mass of *L. latifolium* declines significantly as soil fertility levels decline. In contrast, the growth of *B. tectorum*, which has a dense fibrous root system adapted to explore the soil more efficiently, is not reduced significantly as fertility levels decline. Moreover, the decline in *L. latifolium* above ground mass with decreasing soil fertility levels is steepest when *L. latifolium* is competed with *B. tectorum* in the same pot. Most interestingly, as soil fertility levels decline, the above ground mass of *B. tectorum* increases when either grown by itself or in competition with *L. latifolium*. This finding suggests that as soil fertility declines there is a threshold when more resource efficient plants or plants adapted for resource searching may outcompete *L. latifolium*. This suggestion is supported by shoot/root ratio data. In the most nutrient rich system (the fertile soil in the first growth cycle), *L. latifolium* can allocate more resources to above ground growth. As the nutrient supplying capacity of this fertile soil diminished in the second and third growth cycles, *L. latifolium* must allocate relatively more resources to root growth and the plants are shorter in stature.

Is such greenhouse research scalable to the field? Where we harvested the total plant in the greenhouse, in natural systems a portion of the litter will be decomposed and mineralized thereby maintaining soil fertility levels. In other words, the nutrient threshold limiting *L. latifolium* competiveness may not occur in the field. It is unclear, however, how nutrient cycling will operate in *L. latifolium* dominated systems. The large residual litter debris suggests that a far greater proportion of nutrients will be unavailable in the short term as compared with native systems. There is a paucity of information on nitrogen cycling in *L. latifolium* monocultures. Moreover, we have little information on the importance of shading in maintaining *L. latifolium* populations and whether or not its thick litter layers are detrimental to the germination and establishment of other species or itself. We are presently researching these unknowns. A working hypothesis has been formulated that the industrial age expansion of many weed species is due to corresponding anthropogenic increases in CO_2 (Mayeux *et al.* 1991). The plant biochemical underpinnings of such a hypothesis are manifold; increased atmospheric levels of CO_2 increase the efficiency of water use in all plants and selectively increase photosynthetic rates of plants possessing the C3 pathway – the pathway of most weed species. Furthermore, there is a wealth of empirical evidence to support such a hypothesis (LaMarche *et al.* 1984; Bazzaz and Garbutt 1988; Bazzaz 1990; Polley *et al.* 1993).

The question arises; would *L. latifolium* be an invasive plant at pre-industrial levels of atmospheric CO_2? We suspect not. Disconcertingly, given the projected increases in atmospheric CO_2, we expect an ever increasing succession of invasive plants over the landscape. A dynamic competition will be played out over the earth, a dance of incomprehensible complexity, repleat with multiple interactions and thresholds. A plant that successfully competes today may not waltz as levels of CO_2 increase in the future.

References

Barber, S.A. 1995. Soil Nutrient Bioavailability: a Mechanistic Approach. John Wiley & Sons, New York.

Bazzaz, F.A. and Garbutt, K. 1988. The response of annuals in competitive neighborhoods: Effects of elevated CO_2. Ecology 69: 937-946.

Bazzaz, F.A. 1990. The response of natural ecosystems to the rising global CO_2 levels. Annu. Rev. Ecol. Syst. 21:167-196.

Billings, W.D. 1992. *Bromus tectorum*, a biotic cause of ecosystem impoverishment in the Great Basin. In: Woodwell, G.M. (ed.) The Earth in Transition: Patterns and Processes of Biotic Impoverishment. Cambridge University Press, New York.

Birch, H.F. 1958. The effect of soil drying on humus decomposition and nitrogen. Plant Soil 10:9-31.

Bremner, J.M. and Mulvaney, C.S. 1982. Nitrogen-total. In: Methods of Soil Analysis – part 2. pp. 595-624. Amer. Soc. Agron. Madison, Wisconsin.

El-Ghareeb, R. 1991. Vegetation and soil changes induced by *Mesembryanthemum crystallinum* L. in a Mediterranean desert ecosystem. J. Arid Envir. 20: 321-330.

Grime, J.P. 1974. Plant Strategies and Vegetation Processes. John Wiley & Sons, New York.

Grime, J.P. 1977. Evidence for the existence of three primary strategies in plants and its relevance to ecological and evolutionary theory. Am. Nat. 111: 1169-1194.

Jenny, H. 1941. Factors of Soil Formation. McGraw-Hill, New York.

Keeney, D.R. and Nelson, D.W. 1982. Nitrogen-availability indices. In: Methods of Soil Analysis – part 2. pp. 711-733. Amer. Soc. Agron. Madison, Wisconsin.

Kiegel, G., Tauchnitz, J.G., and Petzold, G. 1988. Microregional differentiation on landfill type Pau/ B. Heicyia 25:272-277.

Kloot, P.M. 1973. Perennial peppercress, a warning. J. Agric., South Australia 76: 72-73.

LaMarche, V.C. Jr., Graybill, D.A., Fritts, H.C., and Rose, M.R. 1984. Increasing atmospheric carbon dioxide: Tree ring evidence for growth enhancement in natural vegetation. Science 225: 1019-1021.

Lovelock, J.E. 1979. Gaia: A New Look at Life on Earth. Oxford University Press. Oxford.

Lyeik, A. 1989. *Lepidium latifolium* L., a sea-shre species in Norway. Blyttia 47:109-113.

Marschner, H. 1986. Mineral Nutrition of Higher Plants. Academic Press, London.

Mayeux, H.S., Johnson, H.B., and Polley, H.W. 1991. Global change and vegetation dynamics. pp. 62-74 In: James, L.F. et al. (eds.), Noxious Range Weeds. Westview Press, Boulder, Colorado.

Miller, G.K., Young, J.A., and Evans, R.A. 1986. Germination of seeds of perennial pepperweed. Weed Sci. 34:252-255.

Nelson, D.W. and Sommers, L.E. 1982. Total carbon, organic carbon, and organic matter. In: Methods of Soil Analysis – part 2. pp. 539-594. Amer. Soc. Agron. Madison, Wisconsin.

Olsen, S.R. and Sommers, L.E. 1982. Phosphorus. In: Methods of Soil Analysis – part 2. pp. 403-430. Amer. Soc. Agron. Madison, Wisconsin.

Polley, H.W., Johnson, H.B., Marino, B.D., and Mayeux, H.S. 1993. Increase in C3 plant water-use efficiency and biomass over glacial to present CO_2 concentrations. Nature 361:61-64.

Radosevich, S.R. and Holt, J.S. 1984. Weed Ecology: Implications for Vegetation Management. John Wiley & Sons, New York.

Richards, L.A. (ed.) 1954. Diagnosis and Improvement of Saline and Alkali Soils. U.S. Salinity Laboratory Staff, U.S. Dept. of Agric. Handbook no. 60.

Soil Survey Staff. 1984. Soil Survey Laboratory Methods and Procedures for Collecting Soil Samples. USDA-SCS Soil Surv. Invest. Rep. no. 1. US Gov. Print. Office, Wash. D.C.

Van Breemen, N. 1993. Soils as Biotic Constructs Favouring Net Primary Productivity. Geoderma 57:183-211.

Young, J.A., Turner, C.E., and James, L.F. 1995. Perennial Pepperweed. Rangelands 17: 121-123.

Young, J.A., Palmquist, D.E., and Wotring, S.O. 1997. Invasive Nature of *Lepidium latifolium*. This volume.

THE DISTRIBUTION, BIOLOGY AND CONTROL OF *SCHINUS TEREBINTHIFOLIUS* IN SOUTHERN FLORIDA, WITH SPECIAL REFERENCE TO EVERGLADES NATIONAL PARK

David T. Jones and Robert F. Doren
South Florida Natural Resources Center, Everglades National Park, 40001 State Road 9336, Homestead, Florida, FL, 33034, USA

Introduction

In many natural areas throughout the United States, exotic pest plants threaten ecosystem function and are causing landscape-level changes to natural systems and processes. Florida and Hawaii, because of their distinctive geographies, climates and histories, have been especially impacted by invasive exotic plants. This situation illustrates what other states may face as exotic pests enter and spread throughout the country and people become increasingly aware of their negative impacts (U.S. Congress Office of Technology Assessment 1993).

Natural disturbances that are a part of the southern Florida environment have provided increased opportunities for exotic plants to become established, an affect amplified by human activities. Some of these species are considered to be the most serious long-term threat to the region's natural ecosystems, particularly the Everglades, capable of altering native plant populations and communities and ecosystem-level properties of areas they invade. The most successful exotic plants of southern Florida – *Melaleuca quinquenervia* (Melaleuca), *Casuarina equisetifolia* (Australian pine), and *Schinus terebinthifolius* (Brazilian pepper) – are so well adapted to an altered habitat that they out-compete native plants and replace entire natural plant communities.

Schinus terebinthifolius (also referred to as Christmas berry and Florida holly) was reportedly introduced into the United States in the 1840s on account of its ornamental qualities (Gordon and Thomas 1994). Morton (1978) attributed its spread in Florida to the popularization of the plant by Dr. George Stone of Punta Gorda: he widely distributed for planting the seedlings he raised in his nursery in 1926. The plant now inhabitats hundreds of thousands of acres in central and southern Florida, including the vast Everglades ecosystem, a sawgrass-dominated natural wetland extending from Lake Okeechobee south into Everglades National Park, a distance of over 100 miles.

Everglades National Park (the "park"), a World Heritage Site and Biosphere Reserve, encompasses more than 1.5 million acres and is the only subtropical wilderness in the continental United States. The principal habitats found within the park include shallow water-marine (40%), saltwater wetland forests (mangroves) and marshes (32%), freshwater marshes and prairies (27%), and upland pine and tropical

Plant Invasions: Studies from North America and Europe, pp. 81–93
edited by J.H. Brock, M. Wade, P. Pyšek and D. Green

hardwood forests (1%). Loope (1992) considered the park among the top U.S. national parks worst affected by exotic pest plants; approximately one-quarter of the 900 plant species reported from the park are exotic in origin (Whiteaker and Doren 1989). *Schinus terebinthifolius* has been observed to invade most types of habitats in the park, both undisturbed and disturbed, including tropical hardwood forests (hammocks), bayhead tree islands, pine rocklands, sawgrass (*Cladium jamaicense*) marshes, *Muhlenbergia* prairies, and coastal mangrove swamps, but not to the degree it colonizes man-influenced features such as canal banks, abandoned farmlands, and roadsides.

Taxonomy and classification

Schinus terebinthifolius belongs to the Anacardiaceae (cashew or sumac family) comprising approximately 850 species in 73 genera (Mabberley 1987). Three genera are indigenous to southeastern United States: *Rhus* (including *Toxicodendron;* sumac, poison ivy), *Metopium* (poison wood), and *Cotinus* (smoke tree) (Brizicky 1962). A number of additional genera have been introduced and are cultivated for their edible fruits or seeds (*Mangifera*, mango; *Pistacia*, pistachio; *Spondias*, hog plum) and ornamental qualities (*Schinus*, pepper tree).

Barkley (1944) divided *Schinus,* a genus of about 30 species, into two subgenera. *Schinus terebinthifolius* (together with about five other species, including California pepper, *Schinus molle*) belongs to subgenus *Euschinus* which is characterized by unarmed shrubs and trees having mostly compound leaves. The remaining species are placed in subgenus *Duvaua*, all of which possess simple leaves and mostly armed branches.

Barkley (1944) recognized five varieties of *S. terebinthifolius*, three of which were known to occur in the United States as introduced taxa at the time of his publication. Of these, var. *terebinthifolius* and var. *raddianus* were reported from Florida. Campbell *et al.* (1980) commented on the possibility of hybridization occurring between these two varieties in the state. He also noted that *S. terebinthifolius* is extremely variable and that many exceptions to the general morphological descriptions can be expected. Because morphological characters often overlap in the field, southern Florida populations have not been adequately characterized or classified to the varietal level.

Schinus terebinthifolius has been referred to by other names in the past, including *S. mucronulata* (in reference to the pointed leaf tip or 'mucro') and *S. antiarthriticus* (in reference to the supposed anti-arthritic action of its resin).

Distribution and ecology

The genus *Schinus* occurs naturally in South America, with one species (*S. molle*) ranging as far north as Mexico. *S. terebinthifolius* is indigenous to subtropical Brazil, Paraguay, and Argentina and has been introduced to various other tropical and subtropical regions of the world including other parts of South America, Central America, the Caribbean Islands, the United States, Mediterranean Europe, northern and southern Africa, China, southern and southeastern Asia, Australia, and the Pacific Islands (Morton 1978; Campbell *et al.* 1980).

In its natural range, *S. terebinthifolius* occurs as scattered individuals in a variety of habitats from sea level to over 700 m elevation (Ewel *et al.* 1982). In southern Florida, it grows on a broad range of moist to mesic sites, often forming nearly monotypic stands. In 1995, it was estimated to cover approximately 800,000 acres from the north rim of Lake Okeechobee south through Everglades National Park (Amy Ferriter, personal communication) (Fig. 1). It thrives particularly on disturbed soils created by natural disruptions, e.g., hurricanes, and is especially invasive in areas affected by human activities, particularly the newly created habitats resulting from agriculture and drainage such as abandoned farmlands, roadsides, and canal banks (Ewel 1986). It does not become established in deeper wetland communities and rarely grows on sites with periods of inundation (hydroperiods) longer than six months. It is typically absent from long hydroperiod marshes and prairies in the park (LaRosa *et al.* 1992).

Investigations in the park on *S. terebinthifolius* invasibility employing seed introduction and seedling transplant experiments in both undisturbed ("native") and disturbed ("successional") plant communities revealed that early successional communities were more susceptible to invasion than older ones, and all successional communities were more susceptible than undisturbed, native communities (Ewel *et al.* 1982). Of three native plant communities investigated, the pinelands were more susceptible to *S. terebinthifolius* seed germination, followed by wet prairies ("glades") and tropical hardwood hammocks (Loope and Dunevitz 1981a).

Concern over the occurrence of *S. terebinthifolius* in salt-tolerant (mangrove) plant communities in southern Florida, especially the park, led Mytinger and Williamson (1987) to investigate the tolerance of the plant to saline conditions. Seed germination and transplanted seedlings did not succeed at salinities of 5 ppm or greater, which would largely exclude it from becoming established in mangrove forest. However, since large areas dominated by the species occur near and in saline coastal areas (a 1987 mapping project revealed an areal extent of about 105,000 acres of *S. terebinthifolius* in the park, occurring in the mangrove forest-freshwater swamp ecotone along the park's western and northwestern coastlines), some other factor must be mitigating (Fig. 1). It is not currently understood how *S. terebinthifolius* and salinity interact in coastal areas.

The stages of secondary plant succession in abandoned, rock-plowed farmlands in southern Florida, leading to nearly pure stands of *S. terebinthifolius*, have been well-documented, particularly in the Hole-in-the-Donut area of Everglades National Park (Loope and Dunevitz 1981a; Ewel *et al.* 1982; Krauss 1987; Doren and Whiteaker 1990). The Hole-in-the-Donut is a major site of *S. terebinthifolius* invasion within the park (Fig. 1). Once a mesic prairie, the area was farmed (and later rock-plowed) for over 35 years until the 1970s when the park acquired it. A monotypic stand of *S. terebinthifolius* has since invaded parts of the site where rock-plowing occurred and now covers approximately 5000 acres. Succession progresses (on sites less than 10 years old) from low *S. terebinthifolius* reproduction and density in a matrix of grasses and herbs, through (on sites 10–20 years old) a stage of rapid reproduction and increased density, to (on sites greater than 20 years old) dense stands of nearly pure *S. terebinthifolius* that are "self-maintaining". The conclusion is that monotypic *S. terebinthifolius* forests represent the final stage in secondary plant succession on abandoned farmlands. The ability of this plant to invade these disturbed habitats may be due to the enhanced conditions created by an altered substrate: the soil is deeper, better drained, better aerated, and possibly more nutrient-rich (Ewel *et al.* 1982).

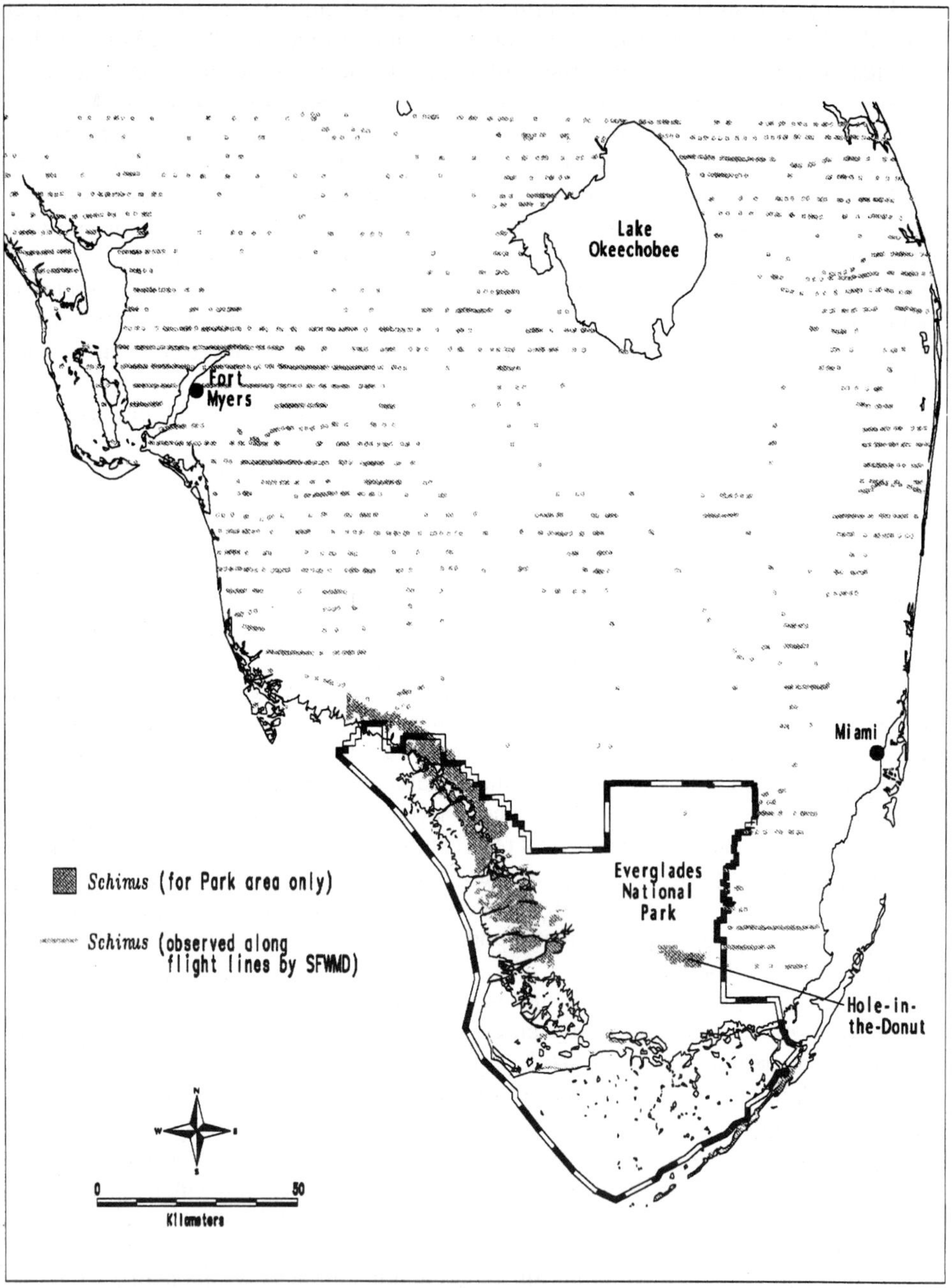

Fig. 1. Distribution of *Schinus terebinthifolius* in Everglades National Park, Florida, USA.

This promotes the growth of mycorrhizal fungi in association with *S. terebinthifolius*, allowing colonization in areas otherwise less suitable (Aziz *et al.* 1995).

The structure of *S. terebinthifolius* stands in the Hole-in-the-Donut was documented by Ewel *et al.* (1982) and Doren and Whiteaker (1990) and revealed stands containing from 200 to more than 2500 *S. terebinthifolius* trees per 2.5 acres. The understory of even the densest stands contain several fern and shrub species. Native (*Myrsine floridana, Persea borbonia, Ilex cassine, Nectandra coriacea)* and exotic *(Psidium guajava,* guava) trees are also present in small numbers. Gogue *et al.* (1974) suggested that *S. terebinthifolius* has the ability to inhibit the growth of competing vegetation through the production of allelopathic substances. Conversely, wax myrtle (*Myrica cerifera*) was shown to be allelopathic and inhibitory to *S. terebinthifolius* germination and seedling establishment (Dunevitz and Ewel 1981). In previously farmed pinelands where wax myrtle has become dominant, *S. terebinthifolius* has been observed to have slower growth rates and poorly developed seedlings. Their reduced vigor under these conditions suggests a possible use of wax myrtle in controlling *S. terebinthifolius* (Dunevitz and Ewel 1981).

Schinus terebinthifolius stands provide relatively poor wildlife habitat. In a study on the utilization of a mature tree stand by the native avifauna, Curnutt (1989) found that avian species diversity and total population density declined when compared to native pinelands and forest-edge habitats. Such results, expected when a species-rich habitat is replaced by one which is biologically less diverse, stress the need to protect native habitats from exotic pest plant encroachment.

Schinus terebinthifolius is not a fire adapted species. Studies conducted in southern California by Nilsen and Muller (1980) indicated that *S. terebinthifolius* seeds are killed by fire; seeds must be introduced to a site after a fire in order to establish seedlings. Fire management practices have so far significantly limited the extent of invasion of limestone rockland pine forests by *S. terebinthifolius* in southern Florida (Loope and Dunevitz 1981b). Because its seedlings require several years after germination in these forests to reach a size capable of surviving fire (about 1 m tall for 80–90% survivorship; % survivorship decreases as seedling height decreases), fires appear to effectively preclude its establishment in pinelands which are regularly burned at less than 5 year intervals (Loope and Dunevitz 1981b). Larger trees usually resprout vigorously from the root-crown after fire. This, in combination with their rapid vegetative growth and ability to out-compete native hardwood species, increases *S. terebinthifolius* dominance of the stand.

Schinus terebinthifolius can largely be excluded from rockland pine forests where adequate fire management programs are carried out. The use of repeated fires as a management tool for controlling successional growth of *S. terebinthifolius* in the pine forests of the park has been investigated by Doren *et al.* 1991. Although repeated burnings may slow down the invasion rate, it does not exclude its establishment; *S. terebinthifolius* invasion progresses with or without the occurrence of fire. Quite the opposite is true for the few remnant pine stands elsewhere in Dade County where lack of such programs has allowed the establishment of *S. terebinthifolius* with a concomitant loss of biological diversity (Doren *et al.*, in prep.).

Schinus terebinthifolius and a closely related species, *S. molle* (California pepper), have both been reported in California as ornamentals for over 100 years. Only the latter species has invaded natural areas in the state and become a pest. Work by Nilsen and Muller (1980a, 1980b) evaluated the difference in invasibility of these two species in California and found that *S. terebinthifolius*'s characteristics of greater drought tolerance, faster growth rate, higher leaf area to weight ratio, higher leaf density, and enhanced ability to extract moisture from the soil should make it a successful invader in California's Mediterannean climate. It has been suggested that seed germination or some other plant characteristic(s) have prevented its naturalization there. It is clear that the underlying causes of exotic plant invasibility are not always evident and that additional research is essential if we are to have any real hope of managing exotic pest plants.

Vegetative and floral morphology

Schinus terebinthifolius is an evergreen shrub or small tree, 3–7 meters tall or more. Its trunk is often multiple-stemmed. Multiple-stemmed trees originate in one of two ways: from sprouting due to damaged trunks and crowns, and from germination of several fruits at the same point, e.g., from seeds dispersed in animal scat. When growing in open areas, the crowns of these trees are broad and rounded and comprise numerous, long, arching, leafy branches which often reach the ground. Ewel *et al.* (1982) noted that these branches do not easily self-prune and "remain attached to the tree, forming an impenetrable tangle that surrounds the tree to ground level". The crowns of trees growing in dense, closed stands, on the other hand, differ in having the foliage concentrated at the top of the canopy, leaving the lower branches relatively leafless.

Each leaf is composed of typically 4 or 6 (or sometimes more) lateral leaflets, arranged in pairs along a narrowly winged leaf axis (rachis), and a single, terminal leaflet (Fig. 2a). The leaf petioles and axes, and expanding leaf blades and shoot apices, are often tinged red. When crushed, the leaves produce a pungent aroma that has been described as "peppery" and "turpentine-like". One or two buds occur in the axils of the leaves (Fig. 2d). If occurring in pairs, the uppermost bud has the potential to develop into a flowering shoot (Tomlinson 1980).

The flowers of *S. terebinthifolius* are produced in showy branched inflorescences which arise from the axils of leaves near the ends of stems (Fig. 2b). *Schinus terebinthifolius* is predominantly dioecious. Ewel *et al.* (1982), however, observed that a small number of trees in a population produce bisexual flowers or are monoecious. Male and female flowers have essentially the same structure: five small green sepals, five small white petals, ten stamens (sterile in female flowers), and a single-chambered ovary (non-functional in male flowers) with three short styles. On female trees, flowering is followed by the production of small, bright red, fleshy, spherical drupes ("berries"), each containing a single seed (Fig. 2c).

Reproductive biology, phenology, and growth

Flowering and fruiting phenomena in *S. terebinthifolius* shows distinct periodicity although they occur sporadically throughout the year (Ewell *et al.* 1982). The main flowering period in southern Florida, September to October, is marked by the production of copious flowers from axillary inflorescences developing at the ends of leafy branches. A second flowering period, March to May, occurs in less than 10%

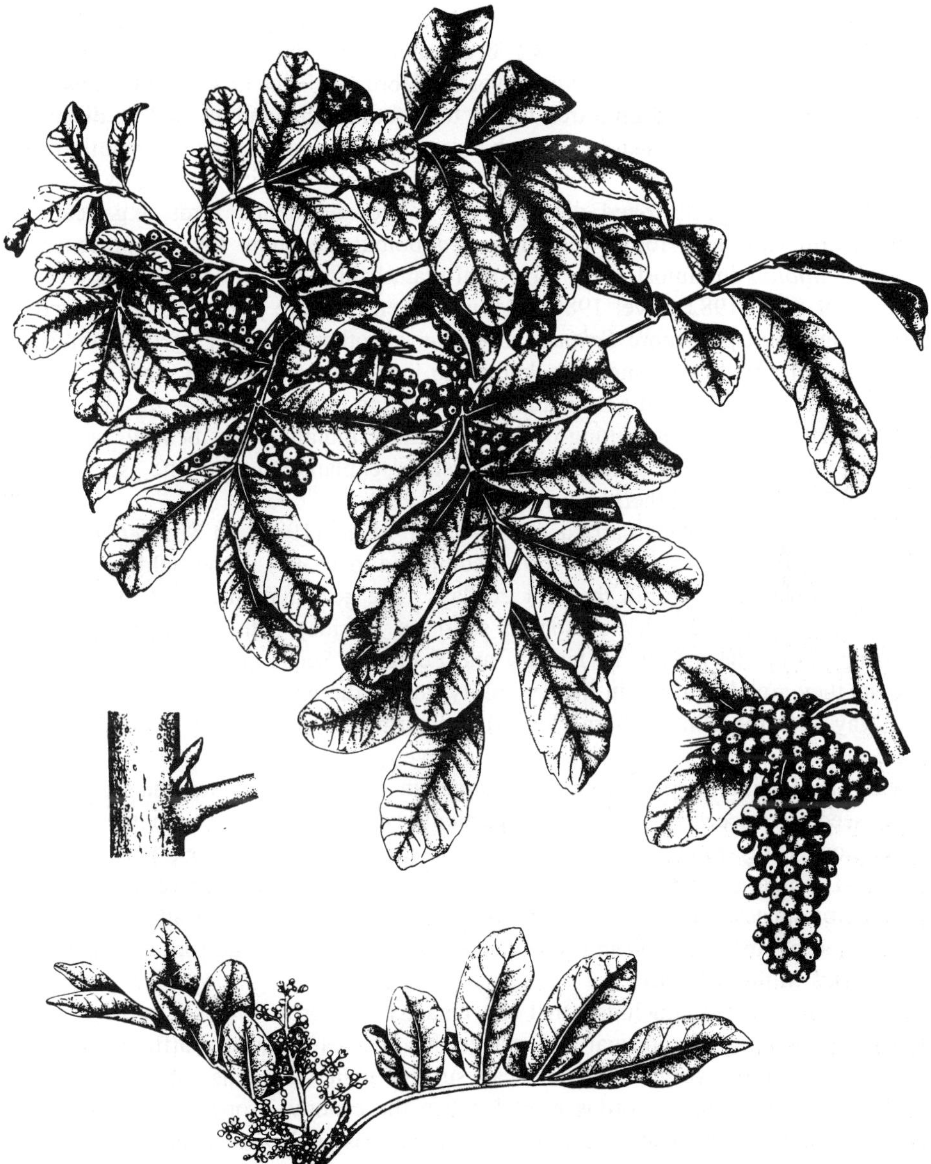

Fig. 2. Foliage morphology of *Schinus terebinthifolius.*

of the population (Ewel *et al.* 1982). Observations by Ewel *et al.* (1982) reveal that the plant is pollinated by diurnal insects, including a number of dipterans (especially a syrphid fly, *Palpada vinetorum*), hymenopterans, and lepidopterans. Pollen availability and nectar secretion in its flowers is apparently timed to maximize pollination success, although Ewel *et al.* (1982) suggested this is unnecessary in southern Florida due to the diversity of local insect pollinators and good fruit set. Plants appear to be out-crossers, although the rare occurrence of fruits developing from unisexual flowers under experimental conditions has not been adequately explained.

Fruit production occurs from November to February. Ewel *et al.* (1982) observed that ripe fruits are retained on a tree for up to eight months, and all will be dispersed before the next flowering season. The abundant and attractive fruits are readily eaten and transported by birds and mammals, with water and gravity serving as less important dispersal agents. Seed dispersal by native and exotic birds, e.g., catbird, mockingbird, American robin, red-whiskered bulbul, accounts for the presence of *S. terebinthifolius* in almost every terrestrial plant habitat in southern Florida (Austin 1978; Ewel *et al.* 1982; Ewel 1986). Robins, when they are present, are believed to consume and transport more *Schinus* seed than all other dispersal agents combined. Raccoons and possibly possums are known to ingest the fruits, their stool providing additional nutrients for seed germination and seedling growth (Ewel *et al.* 1982). The fact that very little else is fruiting during the winter months when *Schinus* seeds are dispersed has been suggested as a possible explanation for the success of *S. terebinthifolius* in southern Florida (Ewel 1986).

Greenhouse experiments carried out by Ewel *et al.* (1982) on *S. terebinthifolius* indicated a germination rate of 20–60% (compared to 1–30% in the field), with most germination occurring within the first 20 days. The germination period ranges from November to April (the dry season in southern Florida) with the highest activity occurring during the months of January and February. Seeds germinating in late fall suggests they are able to retain their viability during the wet season floods (occurring from May to October in southern Florida) and germinate when water levels drop later in the year (Ewel 1979).

Water availability, especially rapid changes in water levels, determines to a great extent seedling success: seedlings are particularly susceptible between the dry and wet seasons (May-June) and during the wet season (July-October), periods when prolonged submergence may result in increased seedling mortality (Ewel *et al.* 1982). The lack of success of this species in southern California has been attributed to the short period of sufficient soil moisture needed for germination and root establishment (Nilsen and Muller 1980a).

Ewel *et al.* (1982) concluded that the tenacity and growth plasticity of *S. terebinthifolius* seedlings is unusual and makes this species especially difficult to manage. Seedlings grow very slowly and can survive under the dense shade of mature stands, while exhibiting vigorous growth when the canopy is opened after a disturbance. In exposed, open areas, such as early successional plant communities, their rates of growth are among the highest: 0.3–0.5 meter per year.

Vegetative growth in *S. terebinthifolius* undergoes a cycle of dormancy in winter when flowering occurs, followed by shoot renewal and extension growth more or less continuously throughout the rest of the year (Tomlinson 1980; Ewel *et al.* 1982). Like many hardwood species, *S. terebinthifolius* has the capability of resprouting from above-ground stems and root crowns under certain conditions such as cutting to a stump, bark girdling, fire (if it girdles a stem), and herbicide application (Woodall 1979). Resprouting is often profuse and the growth rates of the sprouts, which originate from dormant and adventitious buds, are very high. Because of high water tables in southern Florida, the plant's typically shallow root system favors the production of underground root suckers. Root suckers form without evidence of damage to a tree or its root system and can develop into another individual. The clumping of *S. terebinthifolius* often seen during the early stages of invasion may be explained by this suckering mechanism (Woodall 1979).

Ewel (1979) summarized the many characteristics of *S. terebinthifolius* which make it the successful invasive species that it is: (1) fast growth, (2) prolific seed production, (3) near continuous shoot extension and leaf renewal, (4) vigorous resprouting, and (5) tolerance of a wide range of growing conditions. It is unique among weed species in possessing a number of traits more typical of stable, non-invasive plants, including: (1) synchronous flowering and fruiting within a short time period, (2) dioecious trees, (3) insect pollinated flowers, (4) large, animal-dispersed seeds, and (5) shade tolerant seedlings with large cotyledons (an important factor in seedling success).

Toxicity and socio-economic uses

Morton (1969, 1978) reported that *S. terebinthifolius* trees triggered flu-like symptoms, sneezing, sinus congestion, chest pains and acute headache among some people. Its pollen, however, did not appear to be a significant source of irritation or allergies, as it is sticky and not easily carried by wind (Morton 1978). The high concentration of volatile monoterpenes has been suggested as a probable cause of the respiratory problems associated with crushed fruits. Widespread respiratory ailments associated with flowering suggests these compounds may also be produced by the flowers (Lloyd *et al.* 1977). The AMA Handbook of Poisonous and Injurious Plants (Lampe and McCann 1985) reports that the triterpenes found in the fruits can result in irritation of the throat, gastroenteritis, diarrhea, and vomiting in people. Campello and Marsaioli (1974) noted that the ingested fruits have a "paralyzing effect" on birds. Workman (1979) refered to the "hypnotic action" of fruit extracts, containing the triterpenes 'terebinthone' and 'schinol', on chicks and mice.

Like most other members of the Anacardiaceae, *S. terebinthifolius* contains active alkenyl phenols which can cause contact dermatitis and inflammation in sensitive individuals (Lampe and Fagerstrom 1968; Tomlinson 1980). Contact with the "sap" from a cut or bruised tree can result in rash, lesions, oozing sores, severe itching, reddening and swelling (especially of the eyes), and welts (Morton 1978). Grazing animals, such as horses and cattle, are also susceptible to its toxic effects, and ingestion of leaves and/or fruits has been known to be fatal.

Before *S. terebinthifolius* attained its present status as a serious pest plant in southern Florida, it was widely planted along city streets and in home gardens because of its ornamental qualities and for shade (Morton 1969). Its decorative fruiting branches were particularly valued at Christmas and the clusters of fruits were used to make leis and adorn hats (Morton 1978).

As its vernacular name suggests, the dried fruits of *S. terebinthifolius* are used as a spice and sold in the United States as "pink peppercorn". Bell and Taylor (1982) noted that "due to its toxic properties, its use in this way is inappropriate and potentially dangerous". In areas of South America where it occurs naturally, the plant is considered a tonic and astringent. Many different parts of the tree are used for various purposes: stems as the source of a medicinal resin for treating ulcers, respiratory problems, wounds, rheumatism, gout, tumors, diarrhea, skin ailments, and arthritis (Uphof 1968; Campbell *et al.* 1980; Morton 1978); bark for tanning and to preserve fishing lines and nets (Mors and Rizzini 1966; Morton 1978); ignited leaves as "sparklers" for children (Campbell *et al.* 1980); twigs for making toothpicks, wood for making posts, stakes and construction materials, and floral nectar for making honey (Morton 1978) .

Control strategies

The extent of *S. terebinthifolius* infestations in southern Florida typically defy control methods by available resources. Chemical treatments of plants is the preferred method of control, although manual removal of plants can be successful if the entire root system is removed by careful hand pulling. Langeland (1990) has outlined the various herbicide application methods recommended for *S. terebinthifolius*. Many resource managers in southern Florida utilize Garlon 4 applied to the bark around the base of the tree trunk (basal bark method), Garlon 3A applied to the cambium of a freshly cut stump (cut stump method), or Arsenal applied to the foliage (foliar method). All methods require follow-up inspections to determine whether retreatment of surviving trees is necessary. In the case of mechanical removal using heavy equipment, reestablishment of *S. terebinthifolius* from viable seed remaining in the soil seed bank and from sprouting vegetative fragments left in the soil may necessitate follow-up work for several years after initial treatment.

The removal of the dense stand of *S. terebinthifolius* existing in the Hole-in-the-Donut area of Everglades National Park is being accomplished by the total removal of substrate from the site. Experiments carried out independently by the National Park Service and Dade County have provided convincing data that restoration of exotic vegetation-dominated, rock-plowed wetlands is dependent on complete removal of the fundamental disturbance – the artificially created substrate – with concomitant hydrological improvements that favor the reestablishment of native vegetation (Doren *et al.* 1990). The restoration of 5000 acres in the Hole-in-the-Donut (the clearing of the first parcel of land began in November 1996) is funded by the Dade County Freshwater Wetland Mitigation Trust Fund established in 1992 as off-site mitigation to compensate for loss of wetlands in other areas of the county due to growth and development.

Biocontrol on woody exotic plants is an essential tool in the management and control of invasive pest plants. Classical biological control involves utilizing the natural history of the species to understand natural control mechanisms. It comprises moving host-specific natural enemies from the native range of the plant to its introduced range (Habeck 1995). The effectiveness of biological control agents on woody trees and shrubs, as opposed to herbaceous plants, is often questioned. However, the dramatic loss of the American chestnut, and American elm to introduced diseases, and the successful biological control of the weed tree *Cordia curassavica* in Mauritius, Sri Lanka and Malaysia, and other ongoing biocontrol programs on woody weeds, provide clear examples that biocontrol of woody plants is achievable (Bennet and Habeck 1991).

Biological control of *S. terebinthifolius* began in Hawaii during the 1950's. Of 30 insects then identified in Brazil, three species (*Lithraeus atronotatus, Crasimorpha infuscata* and *Episimus utilis*) were introduced into Hawaiian quarantine in 1954 and 1961, and were released in 1963 (Bennet and Habeck 1991). Although two of the species became established, no dramatic control or reduction in spread have been noted to date (Yoshioka and Markin 1988). Biological control work began in Florida in 1986 with initial surveys in southern Brazil for additional insects that might be introduced into Florida. Over 200 insect species have been associated with *S. terebinthifolius*. Two insects have been imported into the United States and are in quarantine for further testing. These include the Brazilian pepper tree thrips, *Liothrips ichini,* and the Brazilian pepper tree sawfly, *Heteroperreyia hubrichi*. Several other insects are being considered for quarantine introductions and are regarded as good candidates for *S. terebinthifolius* biocontrol research (Habeck 1995). Studies to supplement those undertaken on the Hawaiian insects are being done to evaluate their usefulness in southern Florida. Biological control is an important and potent tool in the overall strategy to control of *S. terebinthifolius*, and is considered key to controlling its spread and infestation into areas where other types of control cannot be considered, such as remote mangrove coastal areas in Everglades National Park.

References

Austin, D.F. 1978. Exotic plants and their effects in southeastern Florida. Environ. Conserv. 5: 25-34.

Aziz, T., D.M. Syliva, and R. F. Doren. 1995. Activity and species composition of arbuscular mycorrhizal fungi following soil removal. Ecological Applications 5: 776-784.

Barkley, F.A. 1944. *Schinus* L. Brittonia 5: 160-198.

Bell, C.R. and B.J. Taylor. 1982. Florida Wildflowers and Roadside Plants. Laurel Hill Press, Chapel Hill.

Bennet, F.D. and D.H. Habeck. 1991. Brazilian pepper tree – Prospects for biological control in Florida. In: Center, T.D., R.F. Doren, R.L. Hofstetter, R.L. Myers, and L.D. Whiteaker (eds.). Proceedings of the Symposium on Exotic Pest Plants. Technical Report NPS/NREVER/NRTR-91/06. U.S. Department of the Interior, National Park Service, Washington, D.C. pp. 23-33

Brizicky, G.K. 1962. The genera of Anacardiaceae in the southeastern United States. J. Arn. Arb. 43: 359-375.

Campbell, G.R., J.W. Campbell and A.L. Winterbotham. 1980. The First Fund for Animals, Inc. *Schinus terebinthifolius* Brazil Expedition, July 1980 – Interim Report. [unpublished]

Campello, J.P. and A.J. Marsaioli. 1974. Triterpenes of *Schinus terebinthifolius*. Phytochem. 13: 659-660.

Curnutt, J.L. 1989. Breeding bird use of a mature stand of Brazilian pepper. Fl. Ornithol. Soc. 17: 53-76.

Doren, R.F. and L.D. Whiteaker. 1990. Comparison of economic feasibility of chemical control strategies on differing age and density classes of *Schinus terebinthifolius*. Nat. Areas Journ. 10: 28-34.

Doren, R.F., L.D. Whiteaker, and A.M. LaRosa. 1991. Evaluation of fire as a management tool for controlling *Schinus terebinthifolius* as secondary successional growth on abandoned agricultural land. Environ. Management 15: 121-129.

Doren, R.F., L.D. Whiteaker, G. Molnar, and D. Sylvia. 1990. Restoration of former wetlands within the Hole-in-the-Donut in Everglades National Park. In: Webb, F.J., Jr., (ed.). Proceedings of the Seventh Annual Conference on Wetlands Restoration and Creation. Hillsborough Community College, Tampa.

Dunevitz, V. and J. Ewel. 1981. Allelopathy of wax myrtle (*Myrica cerifera*) on *Schinus terebinthifolius*. Fl. Sci. 44: 13-20.

Ewel, J.J. 1979. Ecology of *Schinus*. In: Workman, R. (ed.). *Schinus* – Technical Proceedings of Techniques for Control of *Schinus* in South Florida: A Workshop for Natural Area Managers. The Sanibel-Captiva Conservation Foundation, Inc., Sanibel. pp. 7-21.

Ewel, J.J. 1986. Invasibility: Lessons from South Florida. In: Mooney, H.A. and J.A. Drake (eds). Ecology of Biological Invasions of North America and Hawaii. pp. 214–230. Springer-Verlag, New York.

Ewel, J.J., D.S. Ojima, D.A. Karl and W.F.DeBusk. 1982. *Schinus* in Successional Ecosystems of Everglades National Park. Report T-676, South Florida Research Center. Everglades National Park, Homestead.

Gogue, G.J., C.J. Hurst, and L. Bancroft. 1974. Growth inhibition by *Schinus terebinthifolius*. Amer. Soc. Hort. Sci. 9: 45.

Gordon, D.R. and K.P. Thomas. 1994. Introduction pathways for invasive non-indigenous plant species. In: Schmitz, D.C. and T.C. Brown (eds.). An Assessment of Invasive Non-Indigenous Species in Florida's Public Lands. Technical Report No. TSS-94-100. pp. 29-43. Florida Department of Environmental Protection, Tallahassee.

Habeck, D.H. 1995. Biological control of *Schinus terebinthifolius*-tree. Florida Naturalist (Spring) pp. 9-11.

Krauss, P. 1987. Old Field Succession in Everglades National Park. Report SFRC-87/03, South Florida Research Center. Everglades National Park, Homestead.

Lampe, K.F. and R. Fagerstrom. 1968. Plant Toxicity and Dermatitis. Williams & Wilkins Co., Baltimore.

Lampe, K.F. and M.A. McCann. 1985. AMA Handbook of Poisonous and Injurious Plants. American Medical Association, Chicago.

Langeland, K. 1990. Exotic Woody Plant Control. Circular 868. Florida Cooperative Extension Service, Institute of Food and Agricultural Sciences. University of Florida, Gainesville.

LaRosa, A.M., R.F. Doren and L. Gunderson. 1992. Alien plant management in Everglades National Park: An historical perspective. In: Stone, C.P., C.W. Smith and J.T. Tunison (eds.). Alien Plant Invasions in Native Ecosystems of Hawaii: Management and Research. pp. 47-63. University of Hawaii Cooperative National Park Resources Studies Unit, Honolulu.

Lloyd, H.A., T.M. Jaouni, S.L. Evans and J.F. Morton. 1977. Terpenes of *Schinus terebinthifolius*. Phytochem. 16: 1301-1302.

Loope, L.L. 1992. An overview of problems with introduced plant species in national parks and biosphere reserves of the United States. In: Stone, C.P., C.W. Smith, and J.T. Tunison (eds.). Alien Plant Invasions in Native Ecosystems of Hawaii: Management and Research. pp. 3-28. University of Hawaii Cooperative National Park Resources Studies Unit, Honolulu.

Loope, L.L. and V.L. Dunevitz. 1981a. Investigations of Early Plant Succession on Abandoned Farmland in Everglades National Park. Report T-644, South Florida Research Center. Everglades National Park, Homestead.

Loope, L.L. and V.L. Dunevitz. 1981b. Impact of Fire Exclusion and Invasion on *Schinus terebinthifolius* on Limestone Rockland Pine Forests of Southeastern Florida. Report T-645, South Florida Research Center. Everglades National Park, Homestead.

Mabberley, D.J. 1987. The Plant Book. Cambridge University Press, Cambridge.

Mors, W.B. and C.T. Rizzini. 1966. Useful Plants of Brazil. Holden-Day Inc., San Francisco.

Morton, J.F. 1969. Some ornamental plants excreting respiratory irritants. Proc. Fl. St. Hort. Soc. 82: 415-421.

Morton, J.F. 1978. *Schinus terebinthifolius* – Its impact on people, animals and the environment. Econ. Bot. 32: 353-359.

Mytinger, L. and G.B. Williamson. 1987. The invasion of *Schinus* into saline communities of Everglades National Park. Fl. Sci. 50: 7-12.

Nilsen, E. and W. Muller. 1980a. A comparison of the relative naturalization ability of two *Schinus* species in southern California. I. Seed germination. Bull. Torrey Bot. Club 107: 51-56.

Nilsen, E. and W. Muller. 1980b. A comparison of the relative naturalization ability of two *Schinus* species (Anacardiaceae) in southern California. II. Seedling establishment. Bull. Torrey Bot. Club 107: 232-237.

Tomlinson, P.B. 1980. The Biology of Trees Native to Tropical Florida. Harvard University Printing Office, Allston.

U.S. Congress Office of Technology Assessment. 1993. Harmful Non-Indigenous Species in the United States. OTA-F-565. U.S. Government Printing Office, Washington, D.C.

Uphof, J.C. 1968. Dictionary of Economic Plants. Verlag von J. Cramer, Lehre.

Whiteaker, L.D. and R.F. Doren. 1989. Exotic Plant Species Management Strategies and List of Exotic Species in Prioritized Categories for Everglades National Park. Report SER-89/04. U.S. Department of the Interior, National Park Service, Atlanta.

Woodall, S. 1979. Physiology of *Schinus*. In: Workman, R. (ed.). *Schinus* – Technical Proceedings of Techniques for Control of *Schinus* in South Florida: A Workshop for Natural Area Managers. pp. 3-6. The Sanibel-Captiva Conservation Foundation, Inc., Sanibel.

Workman, R. 1979. History of *Schinus* in Florida. In: Workman, R. (ed.). *Schinus* – Technical Proceedings of Techniques for Control of *Schinus* in South Florida: A Workshop for Natural Area Managers. The Sanibel-Captiva Conservation Foundation, Inc., Sanibel. pp. 3-6.

Yoshioka, E.R. and G.P. Markin. 1991. Efforts of biological control of Christmas berry (*Schinus terebinthifolius*) in Hawaii. In: Center, T.D., R.F. Doren, R.L. Hofstetter, R.L. Myers, and L.D. Whiteaker (eds.). Proceedings of the Symposium on Exotic Pest Plants. Technical Report NPS/NREVER/NRTR-91/06. U.S. Department of the Interior, National Park Service, Washington, D.C. pp. 377-387

THE STATUS OF *FALLOPIA JAPONICA* (*REYNOUTRIA JAPONICA*; *POLYGONUM CUSPIDATUM*) IN NORTH AMERICA

Leslie A. Seiger
Department of Biology, San Diego State University, San Diego CA 92182-4614, USA, e-mail: lseiger@sunstroke.sdsu.edu

It is perhaps of small comfort to learn that the young shoots of [Fallopia japonica] when cooked like asparagus furnish an estimable article of diet.

J. M. Fogg, Jr.
Weeds of Lawn and Garden

Abstract

The invasive exotic forb *Fallopia japonica* (Japanese knotweed) has become naturalized throughout North America, where its known range extends from Alaska to Georgia. *F. japonica* is a particular problem in riparian areas where its establishment results in loss of wildlife habitat, decreased species diversity, and increased flooding. In North America, *F. japonica* grows in dense stands over four meters high and reproduces solely by an extensive system of rhizomes. Hybrid seeds are produced, but there are no known instances of seedling survival in the wild. Plants disperse when rhizome fragments are washed downstream or transported in soil by humans. Establishment of *F. japonica* appears to be favored in sites with high light and water availability and unrelated to variations in soil texture, pH or nutrients. Established stands are extremely persistent and difficult to eradicate, and supress regrowth of native species.

Introduction

Fallopia japonica (Houtt.) Ronse Decraene (Japanese knotweed), an invasive exotic, is an increasing threat to natural areas in North America, particularly in riparian areas. Previously classified under the synonyms *Polygonum cuspidatum* (Sieb. & Zucc.) and *Reynoutria japonica* Houtt., *F. japonica* is a member of the family Polygonaceae and is closely related to *Fallopia sachalinensis* (Sakhalin knotweed), which also has been introduced to North America (Bailey 1992; Patterson 1977).

F. japonica is an herbaceous perennial which forms large clumps 1–4 meters in height. The stout stems are hollow and bamboo-like, extend erect from the base and are simple or little branched, and glabrous with thinly membranous sheaths. Leaves are broadly ovate, truncate to cuneate at base, abruptly cuspidate, 5–15 cm long, 5–12 cm broad, with petioles 1–3 cm long. Creamy white flowers are 2.5–3 mm long and densely arranged in axillary panicles. Flowers have 3 styles and 8–10 stamens with longitudinally dehiscing anthers. The fruiting calyx is wing-angled and 6–10 mm long. Achenes are shiny black-brown, 2–4 mm long, acutely trigonous, glossy and dark brown. Male flowers have branched panicles on upright racemes with the

Plant Invasions: Studies from North America and Europe, pp. 95–102
edited by J.H. Brock, M. Wade, P. Pyšek and D. Green

distal end of the raceme in the highest position; individual panicles generally point upward. Female flowers are drooping or decumbent with the distal end in the lowest position; individual panicles are not oriented in a particular direction. Both male and female flowers possess vestigial organs of the other sex (Fernald 1950; Muenscher 1955; Locandro 1973; Ohwi 1965). Recent studies suggest that *F. japonica* may be gynodioecious, having both male-sterile and hermaphrodite flowers (Beerling *et al.* 1994). *F. japonica* closely resembles *F. sachalinensis*, which can be distinguished primarily by its larger size, greenish flowers and cordate leaves (Muenscher 1955; Ohwi 1965).

Geographical distribution

F. japonica is native to eastern Asia. It was introduced from Japan to the United Kingdom as an ornamental in 1825, and from there to North America in the late nineteenth century (Conolly 1977; Patterson 1976; Pridham and Bing 1975). *F. japonica* was known to be naturalized in the northeastern USA as early as 1919 (Hitchcock and Standley 1919). Other early records show it was well established in the Washington-Baltimore area by the first half of this century (Norton 1942). Simultaneously, *F. japonica* has expanded its range throughout the United Kingdom and Europe (Conolly 1977; Sukopp and Sukopp 1988; Beerling *et al.* 1994). Beerling *et al.* (1994) recently reviewed its biology in the British Isles and Japan.

F. japonica has received considerably less attention in North America than in the UK, but its present distribution suggests it is already a serious problem and likely to continue its spread (Fig. 1). *F. japonica* has been recorded as naturalized in virtually all states in the eastern half of the USA and in the Pacific northwest, and in Colorado and Utah (Locandro 1978; Patterson 1976; Fernald and Kinsey 1943; Lorenzi and Jeffery 1987; Muenscher 1955; Seymour 1969; Weber 1976; Welsh *et al.* 1987; Steyermark 1963). In Canada, *F. japonica* has been noted from Newfoundland and Nova Scotia to southern Manitoba, and in British Columbia in the Vancouver area (Beerling *et al.* 1994; British Columbia Provincial Museum 1921; Locandro 1978). Its northern limits extend as far north as Alaska (Shetler and Skog 1978). The southern limits of this species appear to be in the states of Georgia and Missouri in the USA (Jones and Coile 1988; Fernald and Kinsey 1943). Its range is not known to extend into Mexico, however agricultural literature from the late nineteenth century indicates that its congener *F. sachalinensis* was considered for use in Mexico as a forage plant and *F. japonica* is sometimes referred to as "Mexican bamboo" (Langman 1964; Everett 1981; Muenscher 1955).

Within the USA, particularly extensive invasions have been noted in western Pennsylvania where *F. japonica* occupies hundreds of hectares of wetlands, streambanks and hillsides and has spread along the banks of the Ohio and Allegheny Rivers and to the islands in these rivers (Wiegman 1992, personal communication). *F. japonica* is also known to be very common in New Jersey, New York, Maryland and Virginia on the east coast, and in Washington in the west. The National Park flora databases list *F. japonica* as present in 25 sites maintained by the US National Park Service, including seven National Parks (US National Park Service 1996). Of these, *F. japonica* has been recorded in Acadia National Park in Maine, Shenandoah National Park in Virginia, Great Smokey Mountains National Park in Tennessee, and

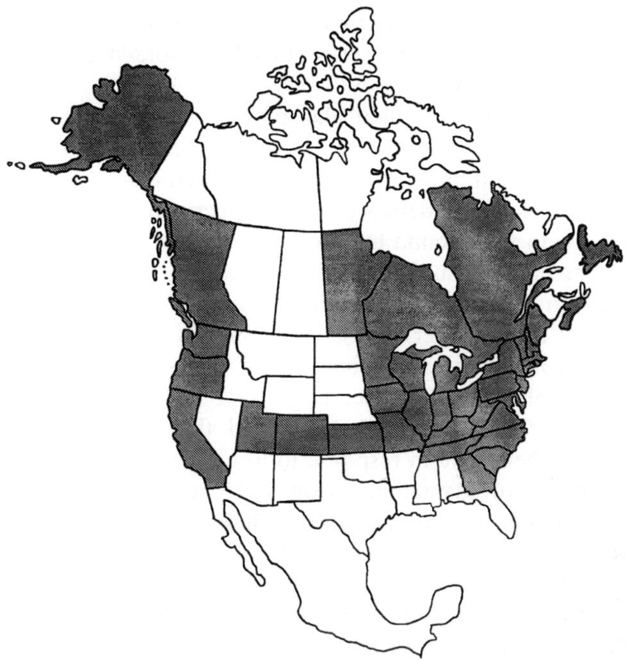

Fig. 1. Distribution of *Fallopia japonica* in North America. Shaded areas indicate states and provinces where *F. japonica* has been recorded as naturalized.

Redwood National Park in northern California. It is considered common in three of the four National Parks in the state of Washington: Mt. Rainier, Olympic and North Cascades National Parks (US National Park Service 1996). As these species lists are provisional, it is likely that it occurs in additional sites.

Habitat

In its native habitat *F. japonica* generally grows on volcanic basalt and riparian washout and is usually found in sunny places on hills and mountains (Hirose 1984; Kanai 1983; Ohwi 1965). In North America, as in Europe and the UK, *F. japonica* grows primarily in riparian habitats and other wet areas, and in areas disturbed by human activity (Beerling 1990; Conolly 1977; Muenscher 1950). In the state of Washington, *F. japonica* is frequently found growing near or along the Cascades (Hitchcock and Cronquist 1973). Based on its European distribution, it appears likely that its distributional limits in Canada are, at this point, restricted by dispersal rather than climatic limits. *F. japonica* grows at latitudes of 63° in Scandinavia and climatic correlations show that it occurs in areas with a heat sum $\geq$ 2505 day-degrees and an absolute minumum temperature of ≥ -30.2 °C (Beerling 1993).

Water and light availability both appear to be of importance in determining the distribution of *F. japonica* where it is introduced. The plant is generally found growing along riverbanks and in other wet areas. Distribution maps from the UK show it is usually associated with regions of high precipitation (Conolly 1977). In the

USA, it has become distributed in virtually all areas with mean annual precipitation greater than 500 mm. In semi-arid areas such as Colorado and Utah, its spread is probably restricted to riparian areas.

Recent research suggests that light also plays a significant role in the ability of *F. japonica* to successfully establish and spread. Reciprocal transplant experiments in mixed hardwood forest in the northeast USA showed that survival and growth were significantly lower in forest understory than on open streambanks. These differences in survival and growth were found to be correlated with light availability in the field (Seiger 1993). Greenhouse studies confirmed that the growth of plants under light levels comparable to that of the forest understory was significantly less than that of plants grown in full light (Seiger 1993). These studies are in accordance with studies from the UK that show growth and abundance of *F. japonica* were depressed on sites with low levels of solar radiation (Beerling 1991; Beerling *et al.* 1994). Further, a physiological study of *F. sachalinensis* showed this congeneric species failed to increase its photosynthetic rate in response to low light intensities and concluded that there was no evidence of adaptation to low light intensity (Patterson *et al.* 1977).

Soil type does not appear to restrict growth of *F. japonica*. In the USA, it has been observed growing in a variety of soil types, including silt, loam, and sand, and in soils with pH ranging from 4.5 to 7.4 (Locandro 1973). In the UK, *F. japonica* has been observed growing in soils ranging in pH from 3.7 to 8.5 and with a wide range of water holding capacity (Beerling *et al.* 1994; Palmer 1990). Analyses of soils from *F. japonica* stands in Wales growing in soils with a broad range of pH, organic matter and nutrients showed no relationship between stand size and vigor and soil characteristics (Palmer 1990). In Japan, *F. japonica* can grow on volcanic soils high in sulfur and having a pH less than 4 and grows in both nutrient poor and nutrient rich sites, with more rapid growth in the latter (Conolly 1977; Hirose 1984). On Mt. Fuji, colonization of bare volcanic desert by *F. japonica* resulted in increased organic nitrogen levels as the plant serves as a nutrient reservoir during primary succession.

Phenology

Data from studies of *F. japonica* in the northeastern USA show that it emerges early in the spring, utilizing its extensive rhizomatous reserves to quickly attain a height which may exceed 4 meters (Seiger, unpublished data). Shoots generally begin to emerge in March or April and growth rates exceeding 8 cm per day have been recorded (Locandro 1973). Full size is attained by the early part of the summer. Flowering occurs for approximately two weeks (personal observation) in late August or early September (Fernald 1950; Muenscher 1955). Leaves senesce following reproduction and stems are killed by the first hard frost, but frequently remain standing into the next growing season.

Reproduction

In Japan, *F. japonica* reproduces both sexually and vegetatively (Maruta 1983; Suzuki 1994). There, sexual reproduction is characterized by high seed production and low seedling survival, with a high probability of survival once plants have developed sufficiently deep roots (Maruta 1976, 1983). Where it has been introduced, however, reproduction appears to be solely vegetative (Beerling *et al.* 1994; Seiger 1993). Plants are pollinated by bees and other insects (Beerling *et al.* 1994; Locandro 1978, personal observation) and populations in the USA may produce copious quantities of seeds (Seiger 1993). However, seedling survival appears to be rare to non-existent (Locandro 1973, 1978; Seiger 1993). In studies of *F. japonica* populations in New Jersey, Locandro (1973) found viable pollen and noted seedling germination in the presence of males. However, he also noted that fertile males were rare and no seedling survival was recorded during five years of observation. In the absence of males, females produced empty achenes.

In contrast to the USA, fruit set is very rare in the UK. This was originally attributed to the rarity of male fertile flowers (Conolly 1977). Bailey (1990) concluded that there are no male fertile forms of *F. japonica* in the UK and that the pollen source is actually a hybrid between *F. japonica* and *F. sachalinensis* (Bailey 1990). *F. japonica* also hybridizes with *Fallopia baldschuanica* (Bailey 1988, 1990). More recent studies of *F. japonica* in the USA showed that plants grown in the laboratory from seeds collected in Washington DC were hybrids (Seiger 1993). These hybrids were morphologically similar to those between *F. japonica* and *F. baldschuanica.* Although there are some reports of seedling establishment in the USA, it is likely that the pollen source is a hybrid, as in the UK, since the North American populations descend from British populations.

Germination rates of the hybrid seeds collected in Washington DC exceeded 50% in two-year old seeds germinated in full light, while germination was suppressed by dark (Seiger 1993). Freshly collected seeds rarely germinated, consistent with literature indicating a dormancy requirement (Beerling *et al.* 1994). Once germinated, seedlings had high mortality in the laboratory (personal observations). No seedling establishment was observed in the wild (Seiger 1993).

Reproduction where *F. japonica* has been introduced is primarily through fragments of the extensive rhizomes which can reach 15–20 meters in length (Conolly 1977; Locandro 1973; Seiger 1993). Plants can regenerate from as little as 7 grams of rhizome (Child *et al.* 1992) and have been observed to regenerate from internode tissue (Locandro 1973). Rhizomes can regenerate when buried up to 1 meter deep and have been observed growing through 5 centimeters of asphalt (Locandro 1978; Pridham and Bing 1975). The ability of rhizomes to generate shoots was found to be affected by the source of rhizome fragments, fragment size and depth planted, the optimal depth being just below the surface (Locandro 1973). Rhizomes are dispersed when fragments from established stands are washed downstream by the current and deposited on banks or when soil contaminated with rhizomes is transported by humans (Conolly 1977; Locandro 1978). In addition, recent research has shown that cut stems are capable of sprouting and establishing new plants (Brock *et al.* 1995).

Effects of invasion by *Fallopia japonica*

F. japonica is primarily a problem in riparian areas. Once established it forms a dense canopy which excludes native vegetation (Gritten 1990; Palmer 1990; Beerling *et al.* 1994; Sukopp and Sukopp 1988). Its extensive rhizomatous reserves allow for early emergence and rapid attainment of great height. This results in shading of other vegetation and prohibits regeneration of other species (Sukopp and Sukopp 1988). In addition, its deep root system gives it an advantage in foraging for nutrients and water (Hirose and Tateno 1984). Further, the highly fibrous stems may alter rates of decomposition and, thus, patterns of nutrient availability.

F. japonica is of little known value for wildlife and its presence reduces wildlife habitat as well as plant species diversity (Palmer 1990; Scott and Mars 1984; Wiegeman 1992, personal communication). The presence of *F. japonica* is also associated with increased flooding because the dense stands reduce water carrying capacity of river and stream channels (Welsh Development Agency 1991). This characteristic may act as a positive feedback switch for further population expansion. Flooding aids in downstream dispersal of rhizomes and, if floods are sufficiently large, the resulting disturbance may increase the area of habitat suitable for *F. japonica*.

Once *F. japonica* has established it is extremely persistent and difficult to eradicate. A number of herbicides have been tested for use against *F. japonica*, the most effective being the persistent herbicide picloram, which is unsafe for use in riparian areas (Ahrens 1975; Beerling 1990; Gritten 1990; Lynn *et al.* 1979; Palmer 1990; Scott and Mars 1984). Nonpersistent herbicides, including glyphosate, are less effective and require repeated treatments (Palmer 1990; Beerling *et al.* 1994). In addition, herbicidal control may be undesirable in conservation areas because of problems associated with lack of specificity, impact on non-target species and persistence in the soil. Mechanical control (pulling or cutting) has been considered unsuccessful (Palmer 1990; Beerling *et al.* 1994). However, some evidence suggests that repeated cutting may reduce rhizomatous reserves, but persistence is necessary (Child *et al.* 1992; Seiger and Merchant, in press). A recent study by Seiger and Merchant (in press) determined that a minimum of four cuts are needed during a single growing season to cause a net depletion of below-ground biomass in small plants.

Though it does not appear to be a serious threat in undisturbed shaded areas such as intact forests, *F. japonica* is likely to continue to expand its range in open mesic habitats, in disturbed areas and, in particular, along riparian corridors. In arid climates, it will probably remain restricted to riparian habitats. However, given the ecological importance of water resources in arid areas and its ability to effectively disperse and establish along waterways, *F. japonica* could become a significant problem in areas with arid climates. There is ample reason for concern that this species will become as problematic in North America as it is in the UK.

References

Ahrens, J.F. 1975. Preliminary results with glyphosate for control of *Polygonum cuspidatum*. Proc. Northeast. Weed Science Soc. 29:326.

Bailey, J.P. 1988. Putative *Reynoutria japonica* Houtt. × *Fallopia baldschuanica* (Regel) Holub hybrids discovered in Britain. Watsonia 17:163-181.

Bailey, J.P. 1990. Breeding behaviour and seed production in Alien Giant Knotweed in the British Isles. In: The Biology and Control of Invasive Plants, pp. 110-120. Proceedings of the Conference of the Industrial Ecology Group of the British Ecological Society. Cardiff.

Bailey, J.P. and C.A. Stace. 1990. Chromosome number, morphology, pairing, and DNA values of species and hybrids in the genus *Fallopia* (*Polygonaceae*). Pl. Syst. Evol. 180:29-52.

Beerling, D.J. 1990. The use of non-persistent herbicides to control riparian stands of Japanese Knotweed (*Reynoutria japonica* Houtt). In: The Biology and Control of Invasive Plants, pp. 121-

129. Proceedings of the Conference of the Industrial Ecology Group of the British Ecological Society. Cardiff.

Beerling, D.J. 1991. The effect of riparian land use on occurrence and abundance of Japanese knotweed *Reynoutria japonica* on selected rivers in South Wales. Biol. Cons. 55:329-337.

Beerling, D.J. 1993. The impact of temperature on the northern distribution limits of the introduced species *Fallopia japonica* and *Impatiens glandifera* in north-west Europe. J. Biogeogr. 20:45-53.

Beerling, D.J., Bailey, J.P. and Conolly, A.P. 1994. *Fallopia japonica* (Houtt.) Ronse Decraene (*Reynoutria japonica* Houtt.; *Polygonum cuspidatum* Seib. & Zucc.). J. Ecol. 82:959-979.

British Columbia Provincial Museum. 1921. A preliminary catalogue of the flora of Vancouver and Queen Charlotte Islands. W. H. Cullin, Victoria, B. C.

Brock, J.H., Child, L.E., De Waal, L.C. and Wade, M. 1995. The invasive nature of *Fallopia japonica* is enhanced by vegetative regeneration from stem tissues. In: P. Pyšek, K. Prach, M. Rejmanek and M. Wade (eds.), Plant Invasions: General Aspects and Special Problems, pp.131-139. SPB Academic Publ., Amsterdam.

Child, L.E., de Waal, L.C. and Wade, P.M. 1992. Control and management of *Reynoutria* species (Knotweed). Aspects App. Biol. 29:295-307.

Conolly, A.P. 1977. The distribution and history in the British Isles of some alien species of *Polygonum* and *Reynoutria*. Watsonia 11:291-311.

Everett, T.H. 1981. The New York Botanical Garden Illustrated Encyclopedia of Horticulture. Garland, New York.

Fernald, M.L. 1950. Gray's Manual of Botany. Edn. 8. American Book Company, N.Y.

Fernald, M.L. and A.C. Kinsey. 1943. Edible Wild Plants of Eastern North America. Idlewild Press, Cornwall-on-Hudson, N.Y.

Gritten, R. 1990. The control of invasive plants in the Snowdonia National Park. In: The Biology and Control of Invasive Plants, pp. 80-85. Proceedings of the Conference of the Industrial Ecology Group of the British Ecological Society. Cardiff.

Hirose, T. 1984. Nitrogen use efficiency in growth of *Polygonum cuspidatum* Sieb. et Zucc. Annals of Botany 54:695-704.

Hirose, T., and Tateno, M. 1984. Soil nitrogen patterns induced by colonization of *Polygonum cuspidatum* on Mt. Fuji. Oecologia 61:218-223.

Hitchcock, A.S. and Standley, P.C. 1919. Flora of the District of Columbia and Vicinity: Contributions from the US National Herbarium Vol. 21. Smithsonian Institution, Washington DC.

Hitchcock, C.L. and Cronquist, A. 1973. Flora of the Pacific Northwest: An Illustrated Manual. Univ. Washington Press, Seattle.

Jones, S. B. Jr. and Coile, N.C. 1988. The distribution of the vascular flora of Georgia. Botany Department, University of Georgia, Athens.

Kanai, H. 1983. Study on the distribution patterns of Japanese plants (5): Accumulation of phytogeographical data of popular plants, points and measures. J. Jap. Bot. 59: 257-269.

Langman, I.K. 1964. A Selected Guide to the Literature on the Flowering plants of Mexico. University of Pennsylvania Press, Philadelphia.

Locandro, R.R. 1973. Reproduction ecology of *Polygonum cuspidatum*. PhD Thesis, Rutgers University.

Locandro, R.R. 1978. Weed watch: Japanese bamboo – 1978. Weeds Today 9:21-22.

Lorenzi, H. J. and Jeffery, L. S. 1987. Weeds of the United States and their Control. Van Nostrand Reinhold, New York.

Lynn, L.B., Rogers, R.A. and Graham, J.C. 1979. Response of woody species to glyphosate in northeastern United States. Proc. Northeast Weed Sci. Soc. 33:336-342.

Maruta, E. 1976. Seedling establishment of *Polygonum cuspidatum* on Mt. Fuji. Jap. J. Ecol. 26:101-105.

Maruta, E. 1983. Growth and survival of current-year seedlings of *Polygonum cuspidatum* at the upper distributional limit on Mt. Fuji. Oecologia 60:316-320.

Muenscher, W.C. 1955. Weeds. Second edition. McMillan Press, New York.

Norton, J. B. S. 1942. Polygonaceae of the Washington-Baltimore Area. Unpublished manuscript. Smithsonian Institution, Washington DC.

Ohwi, J. 1965. Flora of Japan. Smithsonian Institute, Washington, DC.

Palmer, J.P. 1990. Japanese knotweed (*Reynoutria japonica*) in Wales. In: The Biology and Control of Invasive Plants, pp. 80-85. Proceedings of the Conference of the Industrial Ecology Group of the British Ecological Society. Cardiff.

Patterson, D.T. 1976. The history and distribution of five exotic weeds in North Carolina. Castanea 41:177-180.

Patterson, D.T., Longstreth, D.J. and Peet, M.M. 1977. Photosynthetic adaptation to light intensity in Sakhalin knotweed (*Polygonum sachalinense*). Weed Science 25:319-323.
Pridham, A. M. S. and Bing, A. 1975. Japanese-bamboo plants. Plants and Gardens 31: 56-57.
Scott, R., and Mars, R.H. 1984. Impact of Japanese knotweed and methods of control. Aspects of applied biology 5: 291–296. Weed control and vegetation management in forestry and amenity areas.
Seiger, L.A. 1993. The Ecology and Control of *Reynoutria japonica* (*Polygonum cuspidatum*). PhD Thesis, The George Wasington University.
Seiger, L.A. and H.C. Merchant. *In press.* Mechanical control of Japanese knotweed (*Fallopia japonica)*: Effects of cutting regime on rhizomatous reserves. Natural Areas Journal.
Seymour, F.C. 1969. The Flora Of New England; A Manual For The Identification Of All Vascular Plants, Including Ferns And Fern Allies And Flowering Plants Growing Without Cultivation In New England. C. E. Tuttle Co., Rutland,.
Shetler, S. G. and Skog, L.E. (eds.) 1978. A provisional checklist of species for Flora North America. Published for MAB/FNA by Missouri Botanical Garden, St. Louis, Missouri.
Steyermark, J. A. 1963. Flora of Missouri. Iowa State University Press, Ames Iowa.
Sukopp, H. and Sukopp, U. 1988. *Reynoutria japonica* Houtt. in Japan und in Europa. Veröff. Geobot. Inst. ETH, Stiftung Rübel, Zürich 98:354-372.
Suzuki, J. 1994. Growth dynamics of shoot height and foliage structure of a rhizomatous perennial herb, *Polygonum cuspidatum*. Annals of Botany 73:629-638.
US National Park Service. 1996. NPFlora database. gopher://ice.ucdavis.edu/77/NPS/.index/plant lists. Database searched in April, 1996.
Weber, W. A. 1976. Rocky Mountain Flora: a field guide for the identification of the ferns, conifers, and flowering plants of the southern Rocky Mountains from Pikes Peak to Rocky Mountain National Park and from the plains to the Continental Divide. Edt Edn. 5. Colorado Associated University Press, Boulder, Colorado.
Welsh Development Agency. 1991. Guidelines for the Control of Japanese knotweed (*Reynoutria japonica*). Cardiff, Wales.
Welsh, S.L., Atwood, N.D., Goodrich, S. and Higgins (eds.) 1987. A Utah flora. Great Basin Naturalist Memoirs No. 9. Brigham Young University, Provo Utah.

ECOLOGY AND MANAGEMENT OF *ARUNDO DONAX*, AND APPROACHES TO RIPARIAN HABITAT RESTORATION IN SOUTHERN CALIFORNIA

Gary P. Bell
The Nature Conservancy of New Mexico, 212 E. Marcy Street, Suite 200, Santa Fe, NM 87501, USA

Abstract

By far the greatest threat to the dwindling riparian resources of coastal southern California is the alien grass species known as *Arundo donax*. Over the last 25 years the riparian forests of coastal southern California have become infested with *A. donax* which has spread by flood-fragmentation and dispersal of vegetative propagules. *Arundo donax* dramatically alters the ecological/successional processes in riparian systems and ultimately moves most riparian habitats towards pure stands of this alien grass. By current estimates there are tens of thousands of acres of *A. donax* along the major coastal drainage systems of southern California, including the Santa Ana, Santa Margarita, Ventura, Santa Clara, San Diego, and San Luis Rey rivers. The removal of *A. donax* from these systems provides numerous downstream benefits in terms of native species habitat, wildfire protection, water quantity and water quality.

Introduction

Arundo L. is a genus of tall perennial reed-like grasses (Poaceae) with six species native to warmer parts of the Old World. *Arundo donax* L. (giant reed, bamboo reed, giant reed grass, arundo grass, donax cane, giant cane, river cane, bamboo cane, canne de Provence), is the largest member of the genus and is among the largest of the grasses, growing to a height of 8 m (Fig. 1). This species is believed to be native to freshwaters of eastern Asia (Polunin and Huxley 1987), but has been cultivated throughout Asia, southern Europe, north Africa, and the Middle East for thousands of years and has been planted widely in North and South America and Australia in the past century (Perdue 1958; Zohary 1962). It was intentionally introduced to California from the Mediterranean in the 1820's in Los Angeles area as an erosion-control agent in drainage canals, and was also used as thatching for roofs of sheds, barns, and other buildings (Hoshovsky 1987). Subsequent plantings have been made for the production of reeds for a variety of musical instruments including bassoons and bagpipes. Today it is an invasive pest throughout the warmer coastal freshwaters of the United States, from Maryland to northern California.

Arundo donax is a hydrophyte, growing along lakes, streams, drains and other wet sites. It uses prodigious amounts of water, as much as 2,000 L/meter of standing *A. donax*, to supply its incredible rate of growth (Perdue 1958; Iverson 1994). Under optimal conditions it can grow more than 5 cm per day (Perdue 1958). *Arundo donax*

Plant Invasions: Studies from North America and Europe, pp. 103–113
edited by J.H. Brock, M. Wade, P. Pyšek and D. Green

stands are among the most biologically productive of all communities. Under ideal growth conditions they can produce more than 20 tons per hectare above-ground dry mass (Perdue 1958).

Perhaps as much as 90% of the historic riparian habitat in the southern part of California has been lost to agriculture, urban development, flood control, and other human-caused impacts (Jones and Stokes 1987; Katibah 1984). The greatest threat to the remaining riparian corridors today is the invasion of exotic plant species, primarily *Arundo donax*. This alien grass readily invades riparian channels, especially in disturbed areas, is very competitive, difficult to control, and to the best of our knowledge does not provide either food or nesting habitat for native animals. *Arundo* competes with native species such as *Salix* (willows), *Baccharis salicifolia* (mulefat), and *Populus* (cottonwoods) which provide nesting habitat for the federally endangered bird, the least Bell's vireo (*Vireo bellii pusillus*), the federally threatened bird, the willow flycatcher (*Empidonax traillii eximus*) and other native species (Hendricks and Rieger 1989; Franzreb 1989; Zembal 1986, 1990).

Ecological value of native riparian systems

Like most riparian systems, the cottonwood/willow riparian forest is a dynamic community, dependent upon periodic flooding to cycle the community to earlier successional stages (Warner and Hendrix 1985). Periodic floods of large magnitude and migration of the river channel are essential to depositing fresh alluvium where seeds and vegetative propagules of *Baccharis*, *Salix*, and *Populus* can germinate and take root (Gregory et al. 1991; Richter and Richter 1992). Adequate moisture and an absence of subsequent heavy flooding is critical to the survival of the young trees through their first year. As these seedlings mature they increase channel roughness and alter flow during small flood events, increasing sediment deposition (Kondolf 1988; Richter and Richter 1992; Stromberg *et al.* 1993). Sediment deposition builds river terraces and, as they elevate, other plant species colonize resulting in further diversification in the floodplain community (Richter and Richter 1992).

When *Populus/Salix* riparian scrub, which may include such species as *Baccharis salicifolia*, *Vitis californica*, *Rubus ursinus*, and *Urtica dioica* ssp. *holosericea*, reaches four or five years of age, it begins to exhibit the structural diversity required for breeding by the bird, the least Bell's vireo (Franzreb 1989; Hendricks and Rieger 1989). Least Bell's vireo, along with the riparian birds, southwestern willow flycatcher, yellow-breasted chat (*Icteria virens*), yellow warbler (*Denroica petechia*), and many other species may continue to use this diverse community for another ten to twenty years. Gradually the canopy of the maturing *Salix* and *Populus* begins to shade out the diverse understory of vascular plants required by these birds. Older riparian gallery forests will continue to be used by western yellow-billed cuckoo (*Coccyzus americanus occidentalis*), Cooper's hawk (*Accipiter cooperii*), warbling vireo (*Vireo gilvus*) and other species (Zembal 1990; Zembal *et al.* 1985), but as the stand ages the diversity of the flora and fauna within the forest declines. Annual flooding, channel migration, and occasional large flood events maintain this cycle of succession and therefore maintains a mosaic of diverse natural communities (Gregory *et al.* 1991).

Arundo donax as a competitor

Within its introduced range, *A. donax* is an aggressive competitor. *Arundo donax* flowers in late summer with a large, plume-like panicle. Fortunately for California land managers, the seeds produced by *A. donax* in this country are seldom, if ever, fertile. It is not known if this is because of clonal isolation or because of the physiological effects of climate as has been observed in the related *Phragmites communis* (common reed) (Haslam 1958; Rudescu *et al.* 1965). *Arundo donax* is well adapted to the high disturbance dynamics of riparian systems as it spreads vegetatively. Flood events break up clumps of *A. donax* and spread the pieces downstream. Fragmented stem nodes and rhizomes can take root and establish as new plant clones. Thus invasion, spread, and therefore management, of *A. donax* is essentially an intra-basin and downstream phenomenon.

Once established *A. donax* tends to form large, continuous, clonal root masses, sometimes covering several acres, usually at the expense of native riparian vegetation which cannot compete. Root masses, which can become more than a meter thick, stabilize stream banks and terraces (Zohary and Willis 1992), altering flow regimes. *Arundo donax* is also highly flammable throughout most of the year, and the plant appears highly adapted to extreme fire events (Scott 1994). While fire is a natural and beneficial process in many natural communities in southern California it is a largely un-natural and pervasive threat to riparian areas. Natural wild fires usually occur during rare lightening storm events in late fall, winter, and early spring. Under these conditions the moist green vegetation of riparian areas would normally act as a fire break. Human-caused wild fires, in contrast, often occur during the driest months of the year (July through October). Drier conditions in riparian zones at this time of year make them more vulnerable to fire damage. Because *A. donax* is extremely flammable, once established within a riparian area it redirects the history of a site by increasing the probability of the occurrence of wildfire, and increasing the intensity of wildfire once it does occur. If *A. donax* becomes abundant it can effectively change riparian forests from a flood-defined to a fire-defined natural community, as has occurred on the Santa Ana River in Riverside County, California. *Arundo donax* rhizomes respond quickly after fire, sending up new shoots and quickly outgrowing any native species which might have otherwise taken root in a burned site. Fire events thus tend to help push riparian stands in the direction of pure *A. donax*. This results in river corridors dominated by stands of giant reed with little biological diversity.

Arundo donax as habitat

All evidence indicates that *A. donax* provides neither food nor habitat for native species of wildlife. *Arundo donax* stems and leaves contain a wide array of noxious chemicals, including silica (Jackson and Nunez 1964), tri-terpines and sterols (Chandhuri and Ghosal 1970), cardiac glycosides, curare-mimicking indoles (Ghosal *et al.* 1972), hydroxamic acid (Zuñiga *et al.* 1983), and numerous other alkaloids which probably protect it from most native insects and other grazers (Miles *et al.* 1993; Zuñiga *et al.* 1983). Areas taken over by *A. donax* are therefore largely depauperate of wildlife. This also means that native flora and fauna do not offer any

Fig. 1. Arundo donax. Plant × 1/3; spikelet and floret × 3 (from Hitchcock and Chase 1950).

significant control mechanisms for *A. donax*. It is uncertain what the natural controlling mechanisms for this species are in the Old World, although infestations of corn borers (Eizaguirre *et al.* 1990), spider mites (El-Enany 1985) and aphids (Mescheloff and Rosen 1990) have been reported in the Mediterranean. In the United States a number of diseases have been reported on giant reed, including root rot, lesions, crown rust, and stem speckle (USDA 1960), but none seems to have seriously hindered the advance of this weed.

Recent studies by the Santa Ana Watershed Project Authority (Chadwick and Associates 1992) suggest that *A. donax* also lacks the canopy structure necessary to provide significant shading of bank-edge river habitats, resulting in warmer water than would be found with a native gallery forest of *Populus* or *Salix*. As a result, riverine areas dominated by *A. donax* tend to have warmer water temperatures, which results in lower oxygen concentrations and lower diversity of aquatic animals, including fishes (Dunne and Leopold 1978). In the Santa Ana River system this lack of streambank structure and shading has been implicated in the decline of native stream fishes including *Gila orcuttii* (arroyo chub), *Gasterosteus aculatus* (three-spined stickleback), *Rhinichthys osculus* (speckled dace), and *Catostomus santaanae* (Santa Ana sucker). This lack of stream-side canopy structure may also result in increased pH in the shallower sections of the river due to high algal photosynthetic activity. In turn, high pH facilitates the conversion of total ammonia to the toxic un-ionized ammonia form which further degrades water quality for aquatic species and for downstream users (Chadwick and Associates 1992).

Control methods

A suite of methods is needed to control *A. donax* depending upon the presence or absence of native plants, the size of the stand, the amount of biomass which must be dealt with, the terrain, and the season.

The key to effective treatment of established *A. donax* is killing of the root mass. This requires treatment of the plant with systemic herbicide at appropriate times of the year to ensure translocation to the roots. Only one herbicide is currently labeled for wetlands use by the EPA; Rodeo®, a tradename formulation of glyphosate, produced by Monsanto Corporation. Glyphosate is a broad-spectrum herbicide which can be used on *A. donax*, *Tamarix ramosissima* (salt cedar), and most other monocots and dicots. It has proven very effective against *A. donax* (Finn and Minnesang 1990; Jackson 1994; USDA Forest Service 1993). Other herbicides might also be used as labels and conditions allow. Monocot-specific chemicals, such as Fusilade-DX® (fluazapop-butyl) and Post® (Sethoxidan), might be particularly useful for treating *A. donax* in stands with a substantial component of native dicots; however, neither is currently labeled for wetlands use.

The most effective treatment on *A. donax* is the foliar application of a two-to-five percent (2–5%) solution of Rodeo applied post-flowering and pre-dormancy at a rate of 0.5 to 1 l/hectare. During this period of time, usually mid-August to early November, the plants are actively translocating nutrients to the rootmass in preparation for winter dormancy which results in effective translocation of herbicide to the roots. Recent preliminary comparison trials on the Santa Margarita River (Omori 1996) indicate that foliar application during the appropriate season results in almost 100%

control, compared with only 5–50% control using cut-stem treatment. Two to three weeks after foliar treatment the leaves and stalks brown and soften creating an additional advantage in dealing with the biomass: cut green stems might take root if left on damp soil and are very difficult to cut and chip. Treated stems have little or no potential for rooting and are brittle. They may be left intact on the ground or chipped *in situ* for mulch.

Cut-stem treatment requires more time and man-power than foliar spraying and requires careful timing. Cut stems must be treated with concentrated herbicide within one to two minutes in order to ensure tissue uptake (Monsanto 1989). This treatment is also most effective post-flowering. The chief advantage of the cut-stem treatment is that it requires less herbicide that can be more-or-less surgically applied to the stem. Because of its reduced efficacy, and due to the labor required, it is rarely cheaper than foliar spraying except on very small, isolated patches or individual plants.

A popular approach to dealing with *A. donax* has been to cut the stalks and remove the biomass, wait three to six weeks for the plants to grow to about one meter tall, then apply a foliar spray of herbicide solution. The chief advantage of this approach is that less herbicide must be applied to treat the fresh growth compared with tall, established plants, and that coverage is often better because of the shorter and uniform-height plants. However, cutting of the stems may result in the plants returning to growth-phase, drawing nutrients from the rootmass. As a result there is less translocation of herbicide to the roots and less root-kill. Therefore many follow-up treatments must be made which negates any initial savings in herbicide and greatly increases the manpower costs.

Pure stands (>80% canopy cover) of *A. donax* or *T. ramosissima* are most efficiently treated by aerial application of an herbicide concentrate, usually by helicopter. Helicopter application can treat at least 50 hectares per day. Special spray apparatus produces extremely fine droplets (400 microns) of concentrated herbicide which actually reduces herbicide use, minimizes over-spray, and results in greater kill.

In areas where helicopter access is impossible, where *A. donax* makes up the understory, where patches are too small to make aerial application financially efficient, or where weeds are mixed with native plants (<80% cover), herbicides must be applied by hand. Street-vehicles with 400 liter spray tanks are a good alternative where road access is available, but small "quad-runner" vehicles equipped with 60 liter sprayers are the preferred approach where the streambed is not so rocky as to prevent access. Twenty liter backpack sprayers are the final alternative where the vegetation is too dense, or the landscape too rugged for vehicles to be effective.

Methods for vegetation removal include use of prescribed fire, heavy machinery (e.g. bulldozers), handcutting by chainsaw or brushcutter, hydro-axe, chipper, biomass burning or removal by vehicle. Removal of the biomass should only be done where the weed cover is so dense as to prevent recovery by native vegetation after treatment, or where cut vegetation might create a debris-dam hazard during flood events. Prescribed fire, or burning piles of stacked biomass, is the most cost-effective way of removing biomass as long as it does not threaten native vegetation or other resources. Chipping is more costly in terms of equipment and labor, and cut, dried chips pose no threat for regeneration or for forming debris dams. Hauling of biomass by vehicle is extremely expensive and should only be done as a last resort. Most landfills will not accept *A. donax* and those that do will only accept it if cut into short lengths and bagged into plastic trash bags, making the labor costs far too great. The use of heavy machinery such the Hydro-ax® is extremely expensive. The machines are very slow – a Hydro-ax can only cut about 1-5 ha per day.

Riparian restoration and management

One of the prime incentives for riparian habitat restoration has been endangered species recovery, including the federal Endangered Species Act (ESA). The ESA has focused attention on declining species and sought to protect those species in greatest risk by provisions against take (Under the ESA the term "take" means to harass, harm, pursue, hunt, shoot, wound, kill, trap, capture, or collect, or to attempt to engage in any such conduct). Focus of the legislation has been on individual protected species with little attention given to the dynamics of the natural systems of which these species are a part. There are important historical and legislative reasons for this approach. In the 1970's, when the ESA was drafted, ecologists and wildlife managers were highly focused on single species; system-oriented approaches were not widely applied. In addition, it is far easier to attach legal definition to something tangible, such as an individual animal, than it is to the more vague concept of ecological processes (Gregory *et al.* 1991).

The concept of habitat restoration developed in response to the "take" provisions of the ESA as a means of mitigating site-specific damage. While re-vegetation has been carried out in a wide variety of natural community types, its earliest successes and its greatest application has been in mitigation of losses of riparian forests. In southern California, riparian re-vegetation has been pursued as an ever-evolving artform in response to the perceived need for replacement of habitat for the federally and state endangered least Bell's vireo (*Vireo bellii pusillus*) and a suit of other endangered or candidate species including the western yellow-billed cuckoo (*Coccyzus americanus occidentalis*), and the willow flycatcher (*Empidonax traillii*) (Anderson and Miller 1991; Baird and Rieger 1989; Parra-Sjizz 1989; RECON 1988).

It may be argued that the main reason why riparian re-vegetation has received so much attention is because it is so relatively easy to achieve. This ease is a result of the very dynamics of riparian systems – they are high-disturbance systems composed of flood-adapted and resilient species. *Salix, Populus, Baccharis,* and other riparian plant species establish easily by fragmentation in flood events in addition to seeding in flood-washed sediment beds. As a result riparian re-vegetation essentially requires only plant material (cuttings or rooted stock) and water (irrigation). However, such re-vegetation projects can be extremely expensive.

It is also important to recognize that re-vegetation does not necessarily equate with habitat restoration. While the matrix plant species of habitats are relatively easy to establish, the dynamics of native riparian communities are poorly understood. Establishing a *Salix/Populus* stand on a stream-side terrace will probably not provide the community diversity of a natural stand or the dynamic processes required to establish it. While some re-vegetation programs have been successful in terms of establishing a matrix of riparian habitat which is used by some native species, re-vegetating is the not necessarily the best way to create habitat.

The best way to address habitat loss in southern California riparian systems is through a comprehensive program of eradication of *A. donax, T. ramosissima*, and other invasive aliens, and relying on natural physical processes, especially flood dynamics, for the recovery of native natural communities and species. This approach might be just as easily argued for other high disturbance-adapted communities.

This strategy is based upon two of important factors. First, riparian habitats are flood-dynamic communities, dependent upon natural cycles of flood scouring and sediment deposition to create the proper conditions for community establishment (Gregory *et al.* 1991; Richter and Richter 1992; Stromberg *et al.* 1991). The Santa Ana, Santa Margarita, San Luis Rey, and many other southern California streams have all of the factors necessary for the recovery and maintenance of healthy riparian communities and riparian species. These watersheds retain flood regimes sufficient to move and sort sediment and extensive sources of seed and vegetative propagules for *Salix* and other native riparian plants. Second, the only real threats to the integrity of the system are (1) habitat fragmentation by development and (2) introduced exotic species which have altered the successional dynamics and stability of the natural communities. In other words, the native riparian communities of the Santa Ana and other major riparian corridors (and thus riparian-dependent species such as least Bell's vireo) are limited, not by the capacity of the community to regenerate, or the available area of riparian zones, but by the capacity of native species to compete with aggressive invasive exotic species, chiefly *A. donax*.

The majority of the limited resources available for riparian management on these rivers should therefore be directed at managing for the process of riparian systems: removing the key perturbation from the system, thereby allowing natural flood dynamics to operate and the natural communities to recover. Attempts to re-vegetate riparian species in floodplains that retain both native riparian species and flood regimes are redundant, and resources spent to this end are largely wasted. This is not to imply that riparian (and other habitat) re-vegetation efforts should not be applied; however, they should be applied judiciously and only in situations where specific management goals are achieved by carrying out a re-vegetation project (e.g. closing up an important corridor or re-establishing native species in a depauperate watershed). Relying on natural processes for the recovery of the riparian communities has the following major benefits:

a. Cost-effectiveness. Riparian forest restoration is extremely expensive, often on the order of tens of thousands of dollars per hectare. This necessarily limits the size, and therefore the biological value, of any funded restoration project. *Arundo donax* can be removed from most areas of a river for a fraction of the cost of revegetation, opening up areas for natural re-colonization by native riparian species.

b. Biological value. As indicated above, the high cost of re-vegetation limits the size of restoration projects. Additionally, artificially-produced riparian habitat lacks the high stem densities characteristic of naturally regenerating riparian habitat, making the actual biological value of re-vegetated sites questionable. Much higher value may be achieved by removing invasive exotics such as *A. donax* from the system. Areas opened up for recolonization which are subsequently flood-scoured and naturally seeded or "planted" with vegetative propagules spread by the flood are more likely to recover in high stem density habitat.

c. Natural vulnerability. Riparian systems are, by nature, dynamic. The natural flood process that produces the conditions for natural riparian establishment also puts artificially (and naturally) created habitat areas in flood jeopardy. This makes riparian revegetation a high-risk investment of limited resources. Several expensive revegetation projects on the Santa Margarita and Santa Ana Rivers were damaged or lost to flood scouring in January 1993. Some of these areas recovered with high stem-density *Salix* scrub when *A. donax* was controlled. Other sites, without such weed control efforts, succeeded to high density *A. donax* colonies.

Summary

By virtue of its growth characteristics, adaptations to disturbance, especially fire, its lack of natural predators and competitors in North America, and its unsuitability as food or habitat for native wildlife, *Arundo donax* has established itself as one of the primary threats to native riparian habitats in the western United States.

Control and management of *A. donax* within a watershed requires a coordinated, watershed-wide approach. *Arundo donax* should be removed from the watershed beginning in the upper tributaries to prevent reinfestation of treated downstream sites from upstream sources. Removal of *A. donax* requires treatment with systemic herbicides in order to kill the large root mass.

Past practices of riparian restoration have focused on re-vegetation of small sites without consideration of natural riparian processes. Resources should be spent on managing for the natural dynamic processes of these systems on a watershed-wide scale. In coastal southern California the primary perturbation to the natural riparian succession process is invasion by *A. donax*, and its removal from river systems will have a far greater beneficial effect on most riparian species than planting of riparian vegetation.

Acknowledgments

I thank the agencies and representatives of Team Arundo (the Santa Ana River Arundo Management Task Force) and Team Arundo Del Norte for valuable discussion and information that has been incorporated in this paper. Special thanks are due to Cam Barrows, Shelton Douthit, Paul Frandsen, Nelroy Jackson, Dawn Lawson, John Randall, Brian Richter, Eric Stein, Fari Tabatabai, Valerie Vartanian and Dick Zembal for their input.

References

Anderson, B. W. and Miller, E. 1991. Analysis of revegetation on the Kern River Preserve 1986-1990. Prepared for The Nature Conservancy, San Francisco, CA.

Baird, K. and Rieger, J. 1989. A restoration design for least Bell's vireo habitat in San Diego County. USDA Forest Service Gen. Tech. Rep. PSW-110.

Chadwick and Associates. 1992. Santa Ana River use attainability analysis. Volume 2: Aquatic biology, habitat and toxicity analysis. Santa Ana Watershed Project Authority, Riverside, CA

Chandhuri, R.K. and Ghosal, S. 1970. Triterpines and sterols from the leaves of *Arundo donax*. Phytochemistry 9: 1895-1896.

Dunne, T. and Leopold, L.B. 1978. Water in environmental planning. W.H. Freeman and Company, New York.
El-Enany, M.A.M. 1985. Life history studies on *Aponychus solimani* Zaher, Gomaa and El-Enany, with first descriptions of adult male and immature stages. (Acari: Tetranychidae). Zool. Soc. Egypt Bull. 35: 86-91.
Eizaguirre, M, Albajes, R. and Galichet, P.F. 1990. A note on the presence in Catalonia (Spain) of a parasitic system bound to the Tachinid fly *Lydella thompsoni* Hertig, a parasitoid of corn borers. Investigacion Agraria Produccion y Proteccion Vegetales 5: 345-348.
Finn, M. and Minnesang, D. 1990. Control of giant reed grass in a southern California riparian habitat. Restoration & Management Notes 8: 53-54.
Franzreb, K. 1989. Ecology and conservation of the endangered least Bell's vireo. U.S. Fish and Wildlife Service, Biological Report 89(1). 17 pp.
Ghosal, S., Chandhuri, R.K., Cutta, S.K. and Bhattachaupa, S.K. 1972. Occurrence of curarimimetic indoles in the flowers of *Arundo donax*. Planta Med. 21: 22-28.
Gregory, S.V., Swanson, F.J., McKee, W.A. and Cummins, K.W. 1991. An ecosystem perspective of riparian zones. Bioscience 41: 540-551.
Haslam, S.M. 1958. Biological flora of the British Isles: *Phragmites communis* Trin. (*Arundo phragmites* L., ? *Phragmites australis* (Cav.) Trin. ex Steudel). List Br. Vasc. Plants No. 665. J. Ecol. 60: 585-610.
Hendricks, B. and Rieger, J. 1989. Description of nesting habitat for the Least Bell's Vireo. USDA Forest Service Gen. Tech. Rep. PSW-110.
Hitchcock, A.S. and Chase, A. 1950. Manual of the grasses of the United States. Misc. Publ. 200, U.S. Dept. of Agriculture, Washington, D.C. , 1051 pp.
Holland, R.F. 1986. Preliminary descriptions of the terrestrial natural communities of California. Non-game Heritage Division, California Department of Fish and Game, Sacramento, 156 pp.
Hoshovsky, M. 1987. *Arundo donax*. Element Stewardship Abstract. The Nature Conservancy, San Francisco, CA, 10 pp.
Iverson, M.E. 1994. The impact of *Arundo donax* on water resources. In: Jackson, N.E., P. Frandsen, S. Douthit (eds.), November 1993 *Arundo donax* workshop proceedings, pp 19-25. Ontario, CA.
Jackson, G.C. and Nunez, J.R. 1964. Identification of silica present in the giant reed (*Arundo donax* L.). J. Agric. Univ. (Puerto Rico) 48: 60-62.
Jackson, N.E. 1994. Control of *Arundo donax*: techniques and pilot project. In: Jackson, N.E., Frandsen, P. and Douthit, S. (eds.), November 1993 *Arundo donax* Workshop Proceedings, pp 27-33. Ontario, CA.
Jones and Stokes Associates. 1987. Sliding toward extinction: the state of California's natural heritage, 1987. Sacramento, CA.
Katibah, E.F. 1984. A brief history of riparian forests in the Central Valley of California. In: R.E. Warner and K.E. Hendrix (eds.), California riparian systems: ecology, conservation, and productive management, pp. 22-29. University of California Press, Berkeley, CA.
Kondolf, G.M. 1988. Hydrologic and channel stability considerations in stream habitat restoration. In: Berger, J. (ed.), Environmental restoration; science and strategies for restoring the earth: selected papers from Restoring The Earth Conference, pp. 214-227. Island Press, Covelo, CA.
Mescheloff, E. and Rosen, D. 1990. Biosystematics studies on the Aphidiidae of Israel (Hymenoptera: Ichneumonoidea). 3. The genera *Adialytus* and *Lysiphlebus*. Israel J. Entomol. 24: 35-50.
Miles, D.H, Tunsuwan, K., Chittawong, V., Kokpol, U., Choudhary, M.I. and Clardy, J. 1993. Boll weevil antifeedants from *Arundo donax*. Phytochemistry (Oxford): 34: 1277-1279.
Monsanto. 1989. Label for Rodeo® aquatic herbicide. Monsanto Company, St. Louis, MO.
Parra-Sjizz, E.A. 1989. Revegetation in the Sepulveda Wildlife Reserve: a seven year summary. Proceedings of the Society for Ecological Restoration Annual Conference,
Omori, G. 1996. Eradicating the giant reed (*Arundo donax*) in riparian areas of Marine Corps Base, Camp Pendleton, California. Unpublished report to Environmental Security, Marine Corps Base, Camp Pendleton, California and The Nature Conservancy. Agri Chemical & Supply, Inc, Oceanside, CA, 7 pp.
Perdue, R.E. 1958. *Arundo donax* – source of musical reeds and industrial cellulose. Economic Bot. 12: 368-404.
Polunin, O. and Huxley, A. 1987. Flowers of the Mediterranean. Hogarth Press, London.
RECON (Regional Environmental Consultants). 1988. Draft comprehensive species management plan for the least Bell's vireo (*Vireo bellii pusillus*). Prepared for San Diego Association of Governments, San Diego, 212 pp + Append.

Richter, B.D. and Richter, H.E. 1992. Development of groundwater and ecological models for protecting a southwestern riparian system. In: Proceedings, First International Symposium on Groundwater Ecology, Tampa, FL.

Rudescu, L., Niculescu, C. and Chivu, I.P. 1965. Monografia stufului den delta Dunarii. Editura Academiei Republicii Socialiste, România.

Scott, G. 1994. Fire threat from *Arundo donax*. In: Jackson, N.E., P. Frandsen, S. Douthit (eds.), November 1993 *Arundo donax* Workshop Proceedings, pp 17-18, Ontario, CA.

Stromberg, J., Patten, D.T. and Richter, B.C. 1991. Flood flows and dynamics of Sonoran riparian forests. Rivers 2: 221-235.

Stromberg, J., Richter, B.C., Patten, D.T. and Wolden, L.G. 1993. Response of a Sonoran riparian forest to a 10-year return flood. Great Basin Nat. 53: 118-130

USDA 1960. Index of plant diseases in the United States. USDA Crop Research Div., Agricultural Research Service, Handbook 165. US Government Printing Office, Washington, DC.

USDA Forest Service. 1993. Eradication of *Arundo donax*, San Francisquito and Soledad Canyons. Environmental Assessment. USDA Forest Service, Angeles National Forest, CA.

Warner, R.E. and Hendrix, K.M. 1985. Riparian resources of the Central Valley and California Desert. California Department of Fish & Game, Resources Agency, Sacramento, CA.

Zembal, R. 1986. The Least Bell's Vireo in the Prado Basin and environs, 1985. Unpublished report, U.S. Fish and Wildlife service, Laguna Niguel, CA.

Zembal, R. 1990. Riparian habitat and breeding birds along the Santa Margarita and Santa Ana Rivers of southern California. In: A.A.Schoenherr (ed.), Endangered plant communities of southern California, pp. 98-114. Southern California Botanists, Special Publ. No. 3, Fullerton, CA.

Zembal, R., Kramer, K.J. and Bransfield, R.J. 1985. Survey of vegetation and vertebrate fauna in the Prado Basin and the Santa Ana River Canyon, California. U.S. fish and Wildlife Service, Laguna Niguel, CA.

Zohary, M. 1962. Plant life of Palestine. Ronald Press, New York.

Zohary, M. and Willis, A.J. 1992. The vegetation of Egypt. Chapman & Hall, London.

Zúñiga, G.E., Argandoña, V.H., Niemeyer, H.M. and Corcuera, L.J. 1983. Hydroxamic acid content in wild and cultivated Gramineae. Phytochemistry 22: 2665-2668.

A CASE STUDY IN *TAMARIX RAMOSISSIMA* CONTROL: SPRING LAKE, NEW MEXICO

Keith W. Duncan
New Mexico State University Cooperative Extension Service, Artesia, NM 88210, USA

Abstract

Tamarix ramosissima is an exotic phreatophyte which occupies almost a million hectares of riparian areas throughout the western United States. *T. ramosissima* is an aggressive competitor often growing in near monoculture stands and is suspected of lowering water tables thus destroying wetlands and wildlife habitats. A *T. ramosissima* control demonstration trial initiated in August 1989 near Artesia, New Mexico indicated that imazapyr herbicide applied aerially at 1.1 kg/ha provided 95% *T. ramosissima* mortality with one application. Additional data indicate the water table at Spring Lake has risen from a depth of greater than 6.0 m below the soil surface, to the soil surface within 34 months after application. Water returned to the surface of Spring Lake in June 1992 almost three years following herbicide application.

Introduction

Tamarix ramosissima is an introduced phreatophyte in western North America. *Tamarix* occupied well over 600,000 ha of riparian acres in 1965 (Robinson 1965). *T. ramosissima* is a vigorous invader of moist pastures, rangelands and riparian habitats. It is of little forage value, and with the possible exception of doves and bees, it is much less valuable to wildlife than are native riparian plant communities.

Ecology

T. ramosissima was introduced into the eastern U.S. in the early 1800's as an ornamental. By the late 1800's, *T. ramosissima* had escaped cultivation and had become established throughout the U.S., particularly in the arid and semiarid southwest. In the early 1900's government agencies and private landowners began planting *T. ramosissima* for the purpose of stream bank erosion control, such as along the Pecos river in New Mexico.

T. ramosissima produces seed abundantly over a long period, generally April to October. Seeds remain viable for only a few weeks. However, newly produced seeds germinate rapidly, generally in less than 24 hours. Germination is dependent upon saturated soils (Horton *et al.* 1960).

Early seedling growth is very slow. But, once established, the primary root of *T. ramosissima* grows steadily downward with little branching until it reaches the water

Plant Invasions: Studies from North America and Europe, pp. 115–121
edited by J.H. Brock, M. Wade, P. Pyšek and D. Green

table. Secondary branching of the root becomes profuse upon contact with the water (Tomanek and Ziegler 1960). Once established, *T. ramosissima* seedlings grow faster than native plants (Johnson 1986). When mature, *T. ramosissima* roots occupy the capillary zone above the water table with some roots in the zone of saturation (Schopmeyer 1974).

T. ramosissima rapidly dominates an area, out competing native plants for sunlight, moisture, and nutrients. Mature plants can withstand both prolonged droughts and long periods of inundation. *T. ramosissima*'s extensive lateral root system plus its characteristic of dripping salt onto the soil beneath its canopy, makes it extremely competitive with native vegetation. Only xeric species or halophytes can tolerate the understory environment of *T. ramosissima* (Brotherson *et al.* 1984).

Hydrology and habitat impacts

An invasion of a flood plain by *T. ramosissima* usually leads to depletion of streamflow, an increase in the area inundated by floods, and an increase in sediment production (Blackburn *et al.* 1982). *T. ramosissima* has been shown to have an extremely high rate of evapotranspiration. In one experiment, *T. ramosissima* transpired from 0.3 cm to 1.0 cm of water per day and from 1.2 m to 3.1 m of water per year (Davenport *et al.* 1982). Annual water losses may total as much as 2.1 cubic meters/square meter (Carmen and Brotherson 1982). As a result of this high evapotranspiration rate, the water table frequently declines in areas dominated by *T. ramosissima*.

The monotypic stand characteristics of *T. ramosissima* also affect wildlife populations. *T. ramosissima* provides little browse and no seed food source for native wildlife species. Wildlife habitat value of *T. ramosissima* is limited to screening cover for mammals, nesting sites for some birds and a pollen source for bees. In comparing the number of birds in *Populus* (cottonwood), *Salix* (willow) and *Prosopis* (mesquite) to *T. ramosissima* stands, *T. ramosissima* consistently had fewer birds (Cohan *et al.* 1978; Anderson and Ohmart 1977). Engel-Wilson and Ohmart (1978) observed more birds in *Populus, Salix* and *Prosopis* communities than in *T. ramosissima*, even though the native plants covered less than 40 ha of a 20,000 ha study area.

Management and control strategies

In California, one *T. ramosissima* removal project at Eagle Borax Spring in Death Valley National Monument, resulted in the restoration of a marsh and the return of surface water (Neill 1983). *T. ramosissima* apparently invaded the marsh on the westside of the valley floor in the 1940's. By the late 1960's the surface water in the marsh had disappeared, the native grasses and trees were being replaced by *T. ramosissima*, and *Prosopis* trees adjacent to the marsh were slowly losing vigor, owing to the competition for ground water. The Park Service burned the *T. ramosissima* in 1972 then began a cutting and stump treatment operation. By 1982, the *T. ramosissima* was gone ". . . the recovery of the marsh has been rapid and impressive. The surface water has returned" (Neill 1983).

Since its introduction into the U.S., *T. ramosissima* has proven to be a very resilient and difficult-to-control species. *T. ramosissima* is considered to be a fire tolerant plant even though it burns readily, owing to the flammability of the plant material. Fires will burn easily through green or dead *T. ramosissima* and will almost always result in total topkill of the burned stand. However, seldom is mature *T. ramosissima* killed with fire. The rootcrown area of *T. ramosissima* is usually well protected from the hot, but fast moving fire. Resprouts often reach 2 to 3 m in height the first growing season after a fire.

Mechanical control practices have shown only slightly greater success in managing *T. ramosissima*. Mowing or shredding *T. ramosissima* has shown results similar to burning. Bulldozing and root plowing have provided the best mortality results of any mechanical control operation. Root plowing attempts to control the plant by cutting the tap root and bringing the stump out of the soil. In doing so, there is intentional soil disturbance leading to considerable soil movement which covers fragments of stems cut by the machinery, thus permitting rooting and new shoot development (Hollingsworth 1973).

T. ramosissima stem tissue, including root crown material, will sprout vigorously and form new plants if buried or partially buried in warm moist soil (Gary and Horton 1965). Mechanical control by cutting or shredding serves only to break down the plant's brittle stems, leaving the root system to vigorously resprout. The use of the root plow and other heavy equipment as a control for *T. ramosissima* has become less frequent because of the regrowth and lack of permanent control due to the resprouting capabilities of *T. ramosissima* (Hollingsworth 1973). Vegetative propagation and regrowth is probably the prime factor in the ability of *T. ramosissima* to resist efforts of control.

For the most part, the response of *T. ramosissima* to herbicides has been varied, with little satisfactory control except under specific conditions or following several years of repeated treatment. The most successful control methods have been burning in combination with herbicides and/or root plowing in combination with herbicides (Howard *et al.* 1983). Repeated sprayings or spraying after mowing increased *T. ramosissima* mortality. The best results were provided by cut stump treatments or basal bark treatments. However, these operations are labor intensive, and therefore, very expensive. Root plowing followed by herbicide applications provided greater *T. ramosissima* mortality than either treatment used alone. Many of the herbicides historically used for *T. ramosissima* control are no longer manufactured or are not currently approved by the EPA for use where *T. ramosissima* grows.

Material and methods

T. ramosissima growing in two 5.26 ha dry lakes near Artesia, New Mexico were aerially sprayed by a fixed-wing aircraft on August 8, 1989. Imazapyr (2-(4,5-dihydro-4-methyl-4-(1-methylethyl)-5-oxo-1H-imidazol-2-yl)-3-pyridinecarboxylic acid) was applied at 1.1 kg ai/ha in a total volume of 65.4 l/ ha with 0.25% v/v Activator surfactant and 0.25% v/v Nalcotrol. The two dry lakes are approximately 30 m apart and were permanent spring-fed lakes prior to *T. ramosissima* invasion in the late 1960's.

A 5.7 cm diameter hole was hand augered into the bottom of the deeper of the two

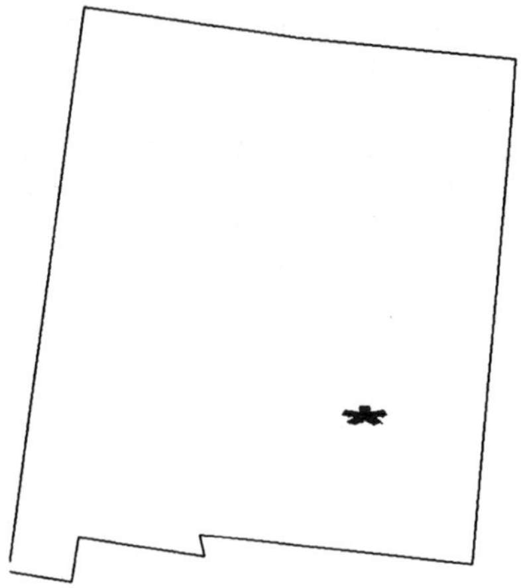

Fig. 1. Map of New Mexico showing approximate location of Artesia, New Mexico.

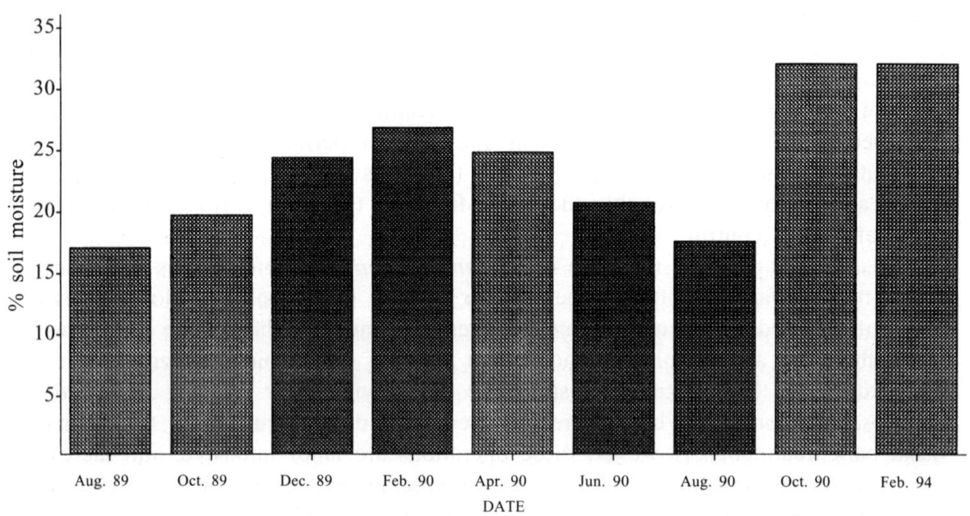

Fig. 2. Soil moisture content at 6.0 m below soil surface following *T. ramossissima* control near Artesia, New Mexico.

lakes on August 15, 1989. The hole was bored to a depth of 5.8 m and a 6.1 m joint of 5.1 cm PVC pipe inserted into the hole. A removable cap was placed over the end of the pipe to prevent moisture or debris from entering the hole from above ground. A soil sample was removed from the bottom of the hole and soil moisture determined gravimetrically. Soil samples were collected and soil moisture determined at approximately 60 day intervals for 12 months.

An attempt was made to collect soil samples in October 1990, 14 months after application. However, the attempt was abandoned when it was determined that the water table had risen to approximately 0.9 m above the bottom of the augered hole. Beginning in October 1990, the depth of the water table was monitored at 30 day intervals using a battery powered water level indicator.

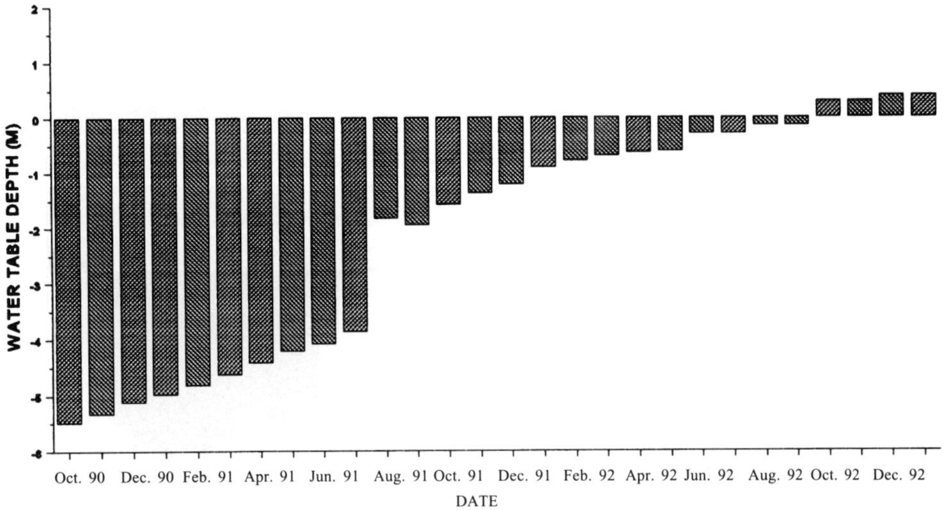

Fig. 3. Water table depths following *T. ramossissima* control near Artesia, New Mexico.

Results and discussion

Soil moisture data indicate that soil moisture at 5.8 m depth was approximately 16.4% on August 15, 1989 (Fig. 2). Soil moisture increased during the winter to a high of 26% in February 1990 then declined throughout the spring and summer. Soil moisture on August 15, 1990 was similar to soil moisture on August 15, 1989 – 16.9% vs. 16.4% respectively. Between August 15, 1990 and October 15, 1990 the water table rose approximately 0.9 m above the bottom of the monitoring well. Depth to the water table has been measured at 30 day intervals since October 1990.

The water table depth monitoring data (Fig. 3) indicate that the water table at Spring Lake rose approximately 0.15 to 0.2 m each month from October 1990 to July 1991. From July to August 1991 the water table rose 2.0 m in conjunction with a large rainfall. The water table dropped slightly from August to September 1991 then rose 0.3 m from September to October 1991 and continued to rise approximately 0.15 to 0.3 m each month until June 1992 when 0.3 m of water was recorded on the surface of Spring Lake. This was the first record of surface water in Spring Lake since 1969. In September 1992 *T. ramosissima* canopy reduction was visually estimated to be 99%. *T. ramosissima* mortality was determined by stem counts to be 95%.

The water depth in Spring Lake increased during winter 1992–1993 to approximately 0.4 m then fell to approximately 0.1 m during summer 1993 (Fig. 4). The water depth rose during winter 1993–1994 but not to the level recorded in winter 1992–1993. In August 1994 the water level in Spring Lake receded below the soil surface after being above ground for 26 months. The water level receded to approximately 1.5 m below the surface in February 1995 and then began a slow rise to reach the surface again in July 1995. Water has remained on the surface of Spring Lake through early October 1995.

Southeastern New Mexico has experienced a mild to severe drought since 1990.

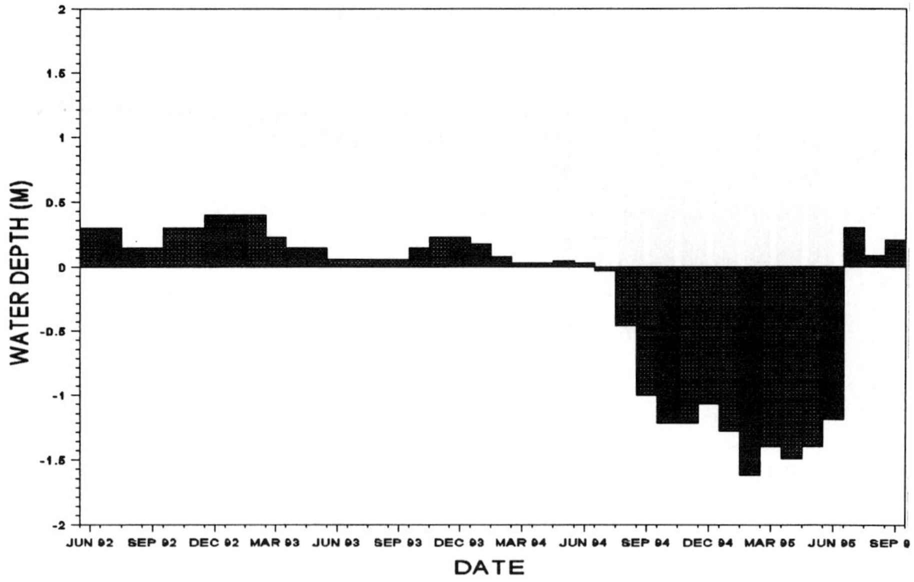

Fig. 4. Water level depths following *T. ramossissima* control near Artesia, New Mexico.

The initial rise in the water table at Spring Lake occurred in conjunction with the beginning of the dry period. The drought has been highlighted by short wet periods, most specifically in July 1991, when the water level in Spring Lake rose 2.0 m in 30 days. However, since 1990 the yearly average precipitation has been 25 to 40% below the long-term average. Summer 1994 was the hottest and driest summer on record for Artesia. The extreme hot and dry conditions are partially responsible for the return below the soil surface of the water level in August. Improving environmental conditions resulted in the water level reaching the soil surface again in July 1995 after 10 months below the surface.

References

Anderson, B.W. and Ohmart, R.D. 1977. Vegetative structure and bird use in the Lower Colorado River Valley. In Importance; preservation and management of riparian habitat; a symposium. USDA, For. Serv. Gen. Tech. Rep. RM-43.

Blackburn, W., Knight, R. W. and Schuster, J. L. 1982. Saltcedar influence on sedimentation in the Brazos River. J. Soil Water Conserv. 37(5): 298-301.

Brotherson, J.D., Carmen, J.G. and Szyska, L.A. 1984. Stem diameter age relationships of *Tamarix ramosissima* in central Utah. J. Range Manage. 37:362-364.

Carmen, J.G. and Brotherson, J.D. 1982. Comparison of sites infested and not infested with Saltcedar (*Tamarix ramosissima*) and Russian Olive (*Elaeagnus augustifolia*). Weed Sci. 30: 360-364.

Cohan, D.R., Anderson, W. and Ohmart, R.D. 1978. Avian population responses to Saltcedar along the Lower Colorado River. USDA For. Serv. Gen. Tech. Rep. WO-12.

Davenport, D.C., Martin, P.E. and Hagar, R.M. 1982. Evapotranspiration from riparian vegetation: water relations and irrecoverable losses for Saltcedar. J. Soil Water Conserv. 37: 233-236.

Engle-Wilson, R.W. and Ohmart, R.D. 1978. Floral and attendant fauna changes on the Lower Rio Grande between Fort Quitman and Presidio, Texas. USDA For. Serv. Gen. Tech. Rep. WO-12.

Gary, H.L. and Horton, J.S. 1965. Some sprouting characteristics of five stamen tamarisk. *U.S. For.* Serv. Res. Note RN-39. 7pp.

Hollingsworth, E.B. 1973. Summary report phreatophyte research Los Lunas, New Mexico 1961-1972. USDA-ARS.

Horton, J.S., Mounts, F.C. and Kraft, J.M. 1960. Seed germination and seedling establishment of phreatophyte species. Rocky Mtn. Forest and Range Experiment Station. U.S. For. Serv. Research Paper 18.

Howard, S.W., Dirar, A.E., Evans, J.O. and Provenza, F.S. 1983. The use of herbicides and/or fire to control Saltcedar (*Tamarisk*). Proc. Western Society of Weed Science 6: 67-72.

Johnson, S. 1986. Can *Tamarisk* be controlled. The Nature Conservancy News, October-November.

Neill, W. 1983. The *Tamarisk* invasion of desert riparian areas. Education Bulletin #83-84. Publication of the Education Foundation of the Desert Protective Council.

Robinson, T.W. 1965. Introduction, spread and real extent of Saltcedar (*Tamarisk*) in the western states. U.S. Geo. Sur. Prof. Paper 491-A. 12 pp.

Schopmeyer, C.S. 1974. Agricultural Handbook No. 450. USFS-USDA, Washington, D.C.

Tomanek, G.W. and Ziegler, R.L. 1960. Ecological studies of salt cedar. Unpub. Report. Division of Biological Sciences. Fort Hays State College, Kansas. 126 pp.

OBSERVATIONS ON FLOWERING AND SEED PRODUCTION IN *HERACLEUM MANTEGAZZIANUM* IN RELATION TO CONTROL

Gordon E.D. Tiley and Bruce Philp
Scottish Agricultural College, Auchincruive, Ayr, KA6 5HW, Scotland

Abstract

Observations on the flowering and fruiting of *Heracleum mantegazzianum* were made in the west of Scotland to quantify its reproductive potential. Over 80,000 flowers were recorded on a single plant but mature fruit numbers were lower. The effects of cutting the flowering stem at different heights, of natural shade and of different population densities on flower and fruit production are reported. The morphology, phenology and hierarchical sequence of inflorescence development is summarised. Apical dominance leads to an earlier and greater development of flowers and larger fruitlets in the terminal umbel of the main stem. Satellite, axillary terminal and lower order branch umbels follow in sequence but the amount of fruitlet development on these depends on the reproductive resources available as determined by environmental and management factors. Increased shade and higher population densities reduced fecundity, as also did cutting the stem at different heights. The results are discussed in relation to control methods.

Introduction

Giant Hogweed, *Heracleum mantegazzianum* Somm. and Lev. (Umbelliferae) was introduced into Britain as a garden ornamental during the 19th century (Stewart 1979). It has subsequently escaped from cultivation and become an invasive and noxious weed. It is particularly abundant along rivers and water courses (Fig. 1), along the sides of roads and railways, on refuse dumps and waste places. (Tiley *et al.* 1996). During the last 30 years it has become noticeably more widespread and invasive in many parts of the United Kingdom (Tiley *et al.* 1996) and also in Europe, e.g. Czech Republic (Pyšek 1991).

H. mantegazzianum is the largest herbaceous plant in Europe, reaching up to 5 m height. It is aesthetically attractive but a number of disadvantages make it undesirable:

1) It is a large, very competitive plant which can shade out the natural flora (Fig. 2). This can thin out the ground vegetation of river banks leading to a risk of erosion in winter (Williamson and Forbes 1982).
2) Dense stands can restrict access in amenity areas and large plants on roadsides can obstruct visibility (Lundström and Darby 1994).
3) *H. mantegazzianum* is a possible alternate host for crop disease (Gray and Noble 1965).
4) Hybridisation occurs with the native common hogweed, *H. sphondylium* (Stewart and Grace 1984).

Plant Invasions: Studies from North America and Europe, pp. 123–137
edited by J.H. Brock, M. Wade, P. Pyšek and D. Green

Fig. 1. River bank invaded by *Heracleum mantegazzianum* on the river Ayr, west Scotland.

5) The sap of *H. mantegazzianum* is poisonous, containing photo-sensitising furano coumarins. In the presence of sunlight the sap causes a skin reaction in humans, varying from a mild reddening to the development of large burn-like blisters (Drever and Hunter 1970). After slow healing, the injured skin may show a permanent pigmentation with a potential for recurrent dermatitis (Anonymous 1980; Camm *et al.* 1976).

Control of local infestations of *H. mantegazzianum* can be straightforward and effective using the correct application of herbicides, cutting or grazing (Dodd *et al.* 1994). Limiting the spread of the species over a wider area requires an organised, and properly executed strategy (Dodd *et al.* 1994), due to the prolific production of seeds which are easily dispersed by water. Seed production has been variously estimated as between 16,000 (Pyšek *et al.* 1995) and 107,000 (Caffrey, J.H., unpublished) per plant. Dispersal by water was studied by Clegg and Grace (1974).

Morphology

Heracleum mantegazzianum develops a perennating taproot and normally remains vegetative for 3 or more years before flowering. The reproducing plant consists of a central main stem with leaves which terminates in an apical (terminal) compound umbel, the largest and first to open. A ring of 5–9 satellite umbels radiate from the base of the peduncle of the terminal umbel. One or two additional satellite umbels (sub-satellites) may originate at points closely below the main satellites. Vigorous or damaged plants develop axillary flowering shoots in 1 or more of the leaf axils from ground level upwards. The axillary branches bear terminal umbels and in turn their own axillary umbels. Satellite and axillary umbels of large vigorous plants may develop diminutive branch umbels subtended by bracts, and occasionally these may bear minute sub umbels, of progressively lower order (Figure 3). While all umbels may flower, only those of the highest order (main terminal, satellite and axillary terminal) develop mature fruitlets (mericarps), depending on plant size, vigour and date of flowering. The flowers are normally hermaphrodite but markedly protandrous. Because of this, lower order umbels may appear to be partly or largely male (Tkatchenko 1989; Ochsmann 1992) *H. mantegazzianum* is amphimictic, insect-pollinated and self-compatible (Tiley *et al.* 1996). Compound umbels contain up to 100 or more rays, each composed of a curving stalk bearing a secondary umbel of pedicelled flowers (Fig. 3).

Phenology

In the west of Scotland over-wintering root crowns develop the first new leaves in January-February, followed, in vegetative plants, by 2–3 large leaves until mid-summer. From July-September there is a slow senescence which is complete by late October. Flowering stems begin elongation in late April. The terminal inflorescence bud appears in late May, followed by the rapid emergence of the terminal umbel and its satellites in sequence. Flowers on the terminal umbel open by early July, the satellites a week later. On branched plants, the axillary terminal umbels flower next with lower order and smaller umbels following in hierarchial order into August or later in favourable summers. Over-wintering seeds germinate in January-March. Established seedlings slowly develop a tap root and a sequence of 3–6 increasingly compound leaves before autumn senescence.

The effects of umbel and leaf removal on floral regeneration were studied by Pyšek *et al.* 1995. Field observations in Scotland indicate reduced numbers of maturing umbels on plants growing in denser shade. Natural stands of flowering *H. mantegazzianum* also exhibit a uniformity of size and population density of approx. 1–2 m^{-2} (Figs. 1 and 2, Gibson *et al.* 1995), implying density-dependent intraspecific competition. The studies reported in this paper centre on the fecundity of flowering and fruiting plants of *H. mantegazzianum*, with observations on how this is influenced by height of cutting, natural shade and experimental differences in population density. The results thus far available from the first phase of the research are presented.

Fig. 2. Flowering and fruiting *Heracleum mantegazzianum* showing the central terminal umbels with early fruit set.

Materials and methods

Location

The studies were carried out at Auchincruive, Ayr in west Scotland (55° 29' N, 4° 34' W), which has a maritime climate of mild winters and cool summers. Mean

Table 1. Rainfall, maximum and minimum temperatures recorded at meteorological station, Auchincruive, Ayr, 1992-1995, 3-monthly values.

Year	Jan.-March	April-June	July-Sept.	Oct.-Dec.	
	Rainfall (mm)				Annual total
1992	321	145	336	329	1131
1993	270	236	189	319	1014
1994	309	137	170	366	982
1995	400	109	208	271	988
	Mean maximum temperature (°C)				Annual mean
1992	8.4	15.5	16.3	8.8	12.2
1993	8.6	13.9	15.9	8.4	11.7
1994	6.9	13.4	17.0	10.9	12.1
1995	7.4	14.4	19.3	9.9	12.7
	Mean minimum temperature (°C)				
1992	3.1	7.7	10.1	2.9	5.9
1993	2.4	6.6	9.3	2.1	5.1
1994	1.9	5.9	10.1	5.8	5.9
1995	1.8	6.7	11.2	4.1	5.9

monthly rainfall and maximum/minimum temperatures, 1992-1995, are given in Table 1. Plants for height of cutting and shade effects were sampled from long-established stands of *H. mantegazzianum* on the banks of the river Ayr. The soil was alluvial sandy clay loam, often overlain with recent sand deposits and with pH 6.3. The shade plants were in deciduous *Quercus petraea-Alnus glutinosa* woodland with an *Agrostis stolonifera* understorey. The height of cutting plants were in mesotrophic grassland (*Agrostis* spp.-*Dactylis glomerata*). Plants for flower and fruitlet counts occurred in both types of habitat. The population density experiment was established in an open field previously used for arable cropping (barley, potatoes). The soil was a sandy clay loam, pH 5.8. A single dressing of $N:P_2O_5:K_2O$ compound fertiliser was applied at the rate of 50:25:25 kg ha^{-1} in the first year to ensure even fertility.

Treatments

Height of cutting

Flowering stems in contiguous plots 3m × 3m within a uniform stand of *H. mantegazzianum* were cut at different levels above or below ground level on 30 June, 1994: ground level, 5 cm below ground level, 15 cm below ground level, 15 cm above ground level, 50 cm above ground level, below the terminal inflorescence bud (inflorescence bud) and untreated (control). Within each cutting treatment 4 plants were sampled at random at the early fruiting stage 7 weeks after cutting.

Effects of shade

Inflorescence production from *H. mantegazzianum* plants growing in heavily shaded deciduous woodland was compared with that from plants in adjacent less shaded areas. Four plants were randomly sampled in each level of shade, 15-20 August 1994. Photosynthetically active radiation (PAR, $\mu mol.m^{-2}s^{-1}$) was measured at 1 m height at a point 1 m from each plant and compared with ambient PAR levels in the open recorded at the same time.

Flower and fruitlet production on large single plants

To provide information on the potential maximum numbers of flowers or mature fruitlets on *H. mantegazzianum*, the apparently largest plants available were sampled for inflorescence size and flower or fruitlet numbers. Flower numbers were recorded in July 1992 and 1993; fruitlet numbers were recorded in a large branched plant in early August 1995. Flower numbers on the terminal umbel were obtained by a total count. Numbers on the remaining satellite and axillary umbels were derived from ray counts per umbel and sample counts of 4 rays. Fruitlet numbers were estimated from fruitlet dry matter (DM) yields per umbel and 100 fruitlet weights, one sample per umbel. This method gave estimates of total matured fruitlet numbers but not of total flower production, which would include infertile and immature flowers. Flower counts were most easily made by counting pedicels in the secondary umbel. In maturing inflorescences one flower equates to two mericarps or fruitlets.

Population density

Experimental plantings of *H. mantegazzianum* with five different densities (Table 6) were established by transplanting seedlings derived from a single plant in the late cotyledon stage in April. Seedlings were planted in rows at the following between-plant spacings: 100 cm, 50 cm, 25 cm, 10 cm, 5 cm. Three rows 4 m long and 50 cm wide were established in each plot except for the 100 cm spacing where row width was 100 cm. Replicates 1 and 2 were planted in 1993 and 3 and 4 in 1994. Vegetative biomass was measured after two seasons growth from 2 × 1 m random lengths from the outer rows per plot. In the third year, when flowering occurred in replicates 1 and 2, two representative plants were sampled per plot at the early fruiting stage to measure vegetative and reproductive parts. Plant dimensions were measured as vertical heights or maximum horizontal diameter. Dry matter weights of plant parts were measured after drying at 105°C for 24 hours. The tap root and its branches were excavated to a depth of 45 cm, washed, weighed and sub sampled for dry weights.

Mean values recorded were compared statistically using standard errors of the mean (SEM) and by analyses of variance. A completely randomised design with sample replication was used in the shade and height of cutting studies. In the population density experiment there was a randomised block design with 5 treatments and 4 replicates; results from 2 replicates are reported here. Treatment and sample replication were both restricted by the large plant size and by the large numbers of flowers and fruitlets produced in *H. mantegazzianum*.

Table 2. Umbel, flower and fruitlet numbers on single large flowering plants of *Heracleum mantegazzianum*. Flowers 1992, 1993; fruitlets 1995. Plant heights: 1992, 283 cm; 1993, 333 cm; 1995, 274 cm. SEM – standard error of the mean.

Umbel type	Umbel number	Rays per umbel Total (mean, SEM)	Flowers or fruitlets per ray (mean, SEM)	Totals (to nearest 100)
			Flowers	
1992				
Terminal	1	122	57.8 (10.2)	7,600
Satellite	5	349 (69.8, 6.9)	50.5 (8.8)	17,600
Axillary terminal	5	451 (75.2, 2.0)	57.2 (10.0)	25,800
Axillary branch	9	677 (84.6, 19.5)	45.1 (11.7)	30,500
Totals	20	1599		81,500
			Flowers	
1993				
Terminal	1	120	67.4, (19.1)	8,000
Satellite	6	523 (87.2, 6.8)	52.6, (17.7)	27,500
Axillary terminal	5	437 (87.4, 19.5)	50.2, (5.8)	21,900
Axillary branch	9	398 (44.2, 7.3)	19.1, (8.9)	7,600
Totals	21	1,478		65,000
			Fruitlets	
1995				
Terminal	1	134	73.8	8,800
Satellite	6	373 (62.2, 2.8)	51.2 (10.5)	19,100
Axillary terminal	5	390 (78.0, 7.5)	63.9 (20.1)	24,900
Axillary branch	14	726 (51.9, 12.1)	0	0
Totals	26	1,623		52,800

Results

Flowering and fruit production

Counts made on single large plants in separate years of total flower numbers (1992, 1993) and total mature fruitlets (1995) are summarised in Table 2, together with umbel and ray numbers. Flower production estimates for the 3 types of umbel (central terminal, satellite and axillary) show that the largest terminal inflorescence produced around 8,000 flowers. In the satellite and axillary categories flower counts per umbel were lower but aggregate totals higher (Table 2). Potential fruitlet production is twice the flower numbers but estimated counts made in 1995 indicate much lower mature fruitlet production. Flower and fruit numbers were higher in the main terminal umbel than in each satellite umbel, and also in axillary terminals compared with lower order branch umbels, reflecting apical dominance in the main and axillary stems. The umbel components: ray length and fruitlet numbers per ray were correlated (Fig. 4), (r^2 = 94.6, P = <0.001), within the terminal inflorescences. Both these parameters were highest in the peripheral rays and decreased towards the centre of

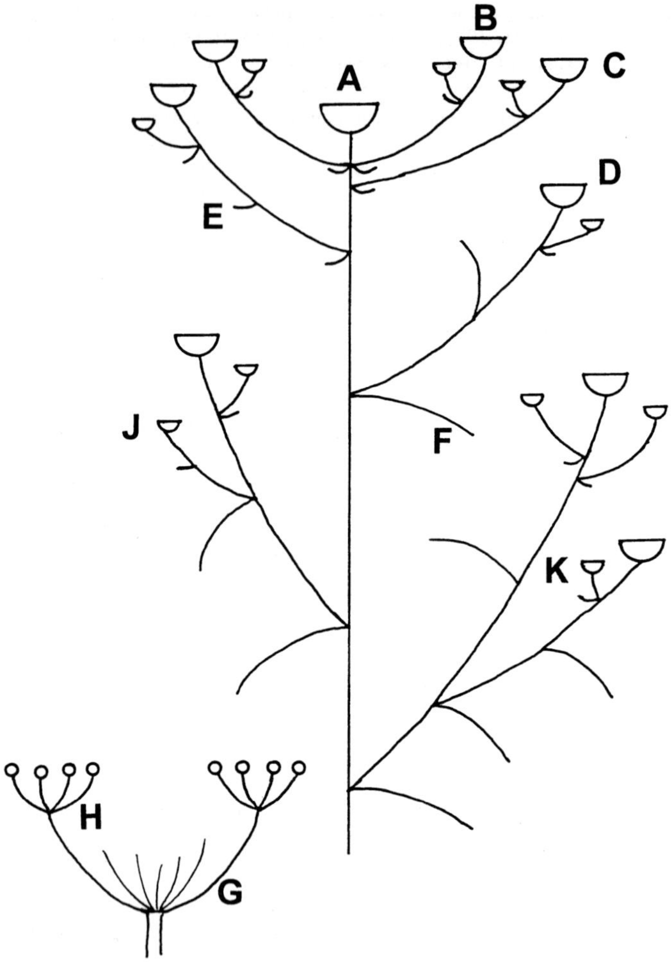

Fig. 3. Diagram of inflorescence hierarchy, position and umbel structure in *Heracleum mantegazzianum*. (A) Terminal umbel; (B) satellite umbel; (C) sub-satellite umbel; (D) axillary terminal umbel; (E) bract; (F) leaf; (G) ray of primary umbel; (H) pedicel with flower in secondary umbel; (J) axillary branch umbel; (K) low order branch umbel. (From Tiley *et al.* 1996).

the umbel (Table 3). Flowers on small, low order umbels e.g. arising below satellite or axillary umbels rarely developed mature fruitlets.

Height of cutting

The effects of cutting flowering plants at different heights above and below ground level are shown in Table 4. Cutting the tap root at 15 cm below ground level (i.e. below the hypocotyl region) prevented further re-growth and effectively killed the plant. Cutting at 5 cm below ground level which is through stem tissue above the hypocotyl region or at ground level (0 cm) allowed re-growth of shoots from axillary buds below ground and resulted in weakened growth and flower production (Table

Table 3. Length of ray stalk and number of *Heracleum mantegazzianum* flowers per secondary umbel in rays developed on different parts of a terminal umbel (Total number of rays = 143). Means of ten rays with SEM. Terminal umbel overall diameter 72 cm.

Location on main umbel	Mean ray length (cm), SEM	Mean flower number per secondary umbel, SEM
Around centre	21.5, 0.9	41.6, 4.4
Intermediate	25.0, 1.7	50.4, 5.9
Intermediate	29.3, 1.5	64.1, 5.6
Intermediate	35.6, 2.1	92.9, 6.2
Around periphery	39.6, 1.4	108.7, 6.3

Table 4. Effect of cutting *Heracleum mantegazzianum* flowering plants at different heights on 30 June. Means of four plants sampled on 14 August. Means followed by the same letter are not significantly different using the least significant difference (lsd) test.

	Regrowth 14 August		
Height of cutting (30 June)	No. of branches	Terminal umbels	Dry weight of umbels (g)
15 cm below ground level	0	0	0
5 cm below ground level	0.8^{ab}	0	0
At ground level	0.3^{a}	1.0^{a}	9.8^{a}
15 cm above ground level	9.8^{c}	3.0^{b}	36.0^{ab}
50 cm above ground level	6.5^{bc}	2.3^{b}	33.8^{ab}
Below central inflorescence bud	4.8^{b}	3.3^{b}	83.5^{b}
Control	6.8^{bc}	1.0^{a}	227.5^{c}
Significance	$P<0.05$	$P<0.01$	$P<0.001$
SEM	3.4	0.9	43

4). Cutting the stem at higher levels gave rise to progressively greater branching with larger numbers and weights of umbels. Total inflorescence yields were substantially less than in the control plants. The seeds produced by the cut plants were shown to be viable.

Effect of increased shade

The effects of two levels of shade, heavy and light, on plant size and inflorescence parameters are shown in Table 5. Associated PAR values, measured in heavy shade were only about one third of those in light shade. The reduced light intensity was associated with reduced height and biomass in the sampled *H. mantegazzianum* plants. In the more heavily shaded plants, size of the terminal umbel was reduced and satellite umbels were largely infertile, axillary branches were absent or feebly developed, and yield and size of mature fruitlets were also lower. Mean DM yield fruitlets (g) was 101 (SEM 30) in deep shade; 208 (68) in light shade; mean of fruitlet weights 1.18 (0.22) in deep shade, 2.25 (0.29) in light shade (1995 data).

Table 5. Effect of shade on plant size and flowering in *Heracleum mantegazzianum*. Means of five plants sampled (a) in deep shade under a summer deciduous tree canopy and (b) in light shade.

	(a) Deep shade	(b) Light shade	SEM	Significance, P-value
PAR values % ambient	12.6	38.9	49.1	NS
Overall height (m)	2.91	3.45	0.56	NS
Dry weights (g)				
Root	121	355	116	<0.05
Stem	129	486	73	<0.001
Leaf	58	115	79	NS
Umbels total	132	401	165	NS
Umbel characteristics – terminal umbel				
Vertical height × diameter (cm)	18 × 68	23 × 75	–	–
Number of rays	105	135	15	<0.05
– satellite umbels				
Fresh weight (kg)	0.48	1.01	0.29	<0.05
Number	4.8	6.6	1	<0.05
Total number of fertile rays	58	759	435	<0.05
Fresh weight (g)	96	771	278	<0.001

Effect of plant population density

A range of population densities from 1–19 plants m^{-2} was attained in the field experiment after plant losses during establishment. The effects of these five population levels on plant size, inflorescence and fruitlet parameters are summarised for replicates 1 and 2 in Table 6.

Flowering

The proportions of plants flowering in the third year at different population densities are given in Table 6. Nearly all plants flowered at the lowest density (100 cm × 100 cm) with a progressive reduction in the percentage of plants flowering as population density increased ($P < 0.05$).

Plant size

Stem diameter at the base of the flowering stem decreased ($P<0.05$) with increase of population density. Overall plant heights were not affected but axillary and satellite branching was reduced as population density increased (Table 6).

Inflorescence production

Total numbers, sizes and weights of satellite and axillary umbels per plant were reduced with increases in population density and total umbel weight was reduced significantly ($P < 0.05$). Terminal umbel weight was not reduced significantly. Ray numbers of satellite and axillary umbels decreased with increasing plant population and floral sterility increased (Table 6). Ray numbers on the terminal umbel decreased non-significantly when population density increased.

Table 6. Effects of different plant population densities on plant size, flowering percentage, umbel and fruitlet parameters in flowering *H. mantegazzianum*; means per treatment, P-value and SEM.

	Population density, plants m^{-2}						Significance, P-value
	1	4	6.5	17.5	19	SEM	
Height (cm)	199	195	209	200	216	7	NS
Plants flowering (%)	83	57	66	33	20	12	<0.05
Stem diameter (cm)	9.5	8.0	6.8	6.3	5.5	0.8	<0.05
Umbels, per plant							
Dry wt (g) – total	188	87	86	53	67	28	<0.05
– terminal	53	37	53	36	41	19	NS
Number – terminal	1	1	1	1	1	-	-
– satellite	11	8.3	7.3	6.3	8.8	1.4	NS
– axillary terminal	3.8	3.3	1.0	1.3	0.3	1.2	NS
Rays, per plant							
Number – terminal	127	113	113	106	108	13	NS
– satellite	849	501	453	305	461	92	<0.05
– axillary terminal	251	172	54	56	27	73	NS
Fruitlets, per plant							
DM yield (g) – terminal	148	118	132	90	92	36	NS
– satellite	220	59	29	19	5	58	NS
– axillary terminal	70	6	0	0	0	-	-
DM yield per umbel (g)							
– terminal	148	118	132	90	92	-	-
– satellite	20.0	7.1	4.0	3.0	0.6	-	-
– axillary terminal	18.4	1.8	0	0	0	-	-
Number – terminal	7100	6300	8300	6100	6700	2000	NS
– satellite	16700	4000	3900	1000	100	3100	<0.05
– axillary terminal	15800	300	0	0	0	-	-
Weight per 100 fruitlets (g)							
– terminal	1.94	1.66	1.52	1.35	1.24	0.14	<0.05
– satellite	0.97	0.66	0.65	0.72	0.50	0.10	<0.01
– axillary terminal	1.00	0.89	-	-	-	-	-

Fruitlet yields, number and size

Yields, estimated numbers and unit weights of fruitlets at different population densities are given in Table 6. Yields and numbers of mature fruitlets on the satellite and axillary umbels decreased markedly with populations above 1 plant m^{-2}. Fruitlet production from satellite and axillary umbels was virtually nil above 6.5 and 4 plants m^{-2}, respectively. Individual fruitlet weights were significantly reduced on both terminal and satellite umbels with increased population density. Data for mean yields of fruitlets per umbel (Table 6) indicate the relative importance of the 3 main types of umbel: terminal, satellite and axillary terminal in reproduction and the decreased role of the satellite and axillary umbels at higher population densities.

Discussion

Flowering in *H. mantegazzianum* progresses in a hierarchical sequence governed by apical dominance of the central terminal umbel, followed by its satellites, the axillary terminal umbels and then later by small, low order umbels from the stems of all but the main terminal umbel (Fig. 1, Shumova 1973). Size of umbel and its components are largely morphogenetic, but degree of fruitlet maturation and fruitlet yields (numbers and size) appear to depend on earliness of flowering, a determinant of reproductive sink strength (Lee 1988). Mature fruitlet production is normally from the terminal and satellite umbels, with additions from axillary inflorescences on more vigorously branched plants. Late-formed low order umbels rarely produce mature fruitlets. Variable proportions of later formed (Stewart and Grace 1984) or tertiary (Ochsmann 1992) umbels are reported to bear male flowers. It is uncertain if this is a physiological response to limitations of reproductive resources. Overinitiation of inflorescences and flowers (Lee 1988) is commonly observed in large *H. mantegazzianum* plants, which continue developing small and depauperate axillary umbels until senescence in the autumn.

Interruption of the normal pattern of flowering, for example, by cutting the main flowering stem and thus removing the main inflorescences and high order umbels results in a stimulation of flowering in the regenerating axillary branches (Pyšek *et al.* 1995), who found that after removal of the terminal umbel in June, compensatory axillary branching resulted in similar total numbers of fruitlets with similar unit size. Removal of all flowers and/or leaves resulted in a reduction of fruitlet numbers and size. In Ireland, Caffrey (unpublished) recorded reduced plant height and seed numbers produced by the regeneration of plants cut to ground level. Effects were much greater when cutting was made late (May) in the growing season. The vigour of regenerating branches and inflorescence yields in the present study varied inversely with the amount of stem removed in June. Similar regeneration of axillary flowering branches has been observed from the base of the plants damaged and incompletely killed by herbicides. Both cutting and herbicide damage can lead to extended perenniality of the rootstock where damaged plants prevented from complete flowering survive for further flowering and fruiting in the following year.

Shade

In its natural habit in the Caucasus mountains, *H. mantegazzianum* occurs in partially shaded locations within glades and at the edges of forests. It is common in similar lightly wooded sites near rivers in Britain (Tiley and Philp 1992), but absent from heavily wooded and very shaded communities. Ochsmann (1992) recorded higher numbers of *H. mantegazzianum* plants in sunny and partly shaded areas than in fully shaded situations. The weakening effect of increased shade on growth and reproduction recorded here confirms the contrasts visible between single-stemmed plants in shade and the robust multi-branched flowering plants growing fully in the open at field edges.

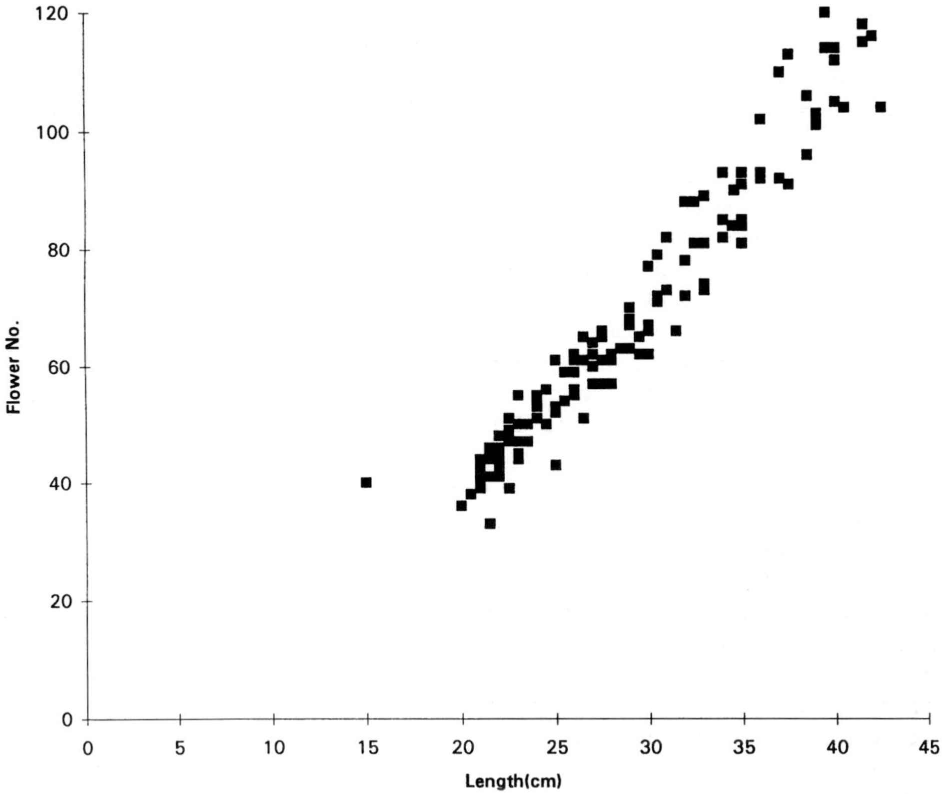

Fig. 4. Relationship of number of flowers on the secondary umbel to ray length in *Heracleum mantegazzianum*. Flower number = – 31.67 + 3.45 ray length, r^2 = 94.6.

Population density

The even-aged stand of *H. mantegazzianum* plants artificially established in the population density trial is only likely to occur naturally in primary colonisation derived from a single seeding plant. The effects of competition from population density on reproduction were reviewed by Weiner (1988), quoting studies where density stress delayed the flowering of biennials. *H. mantegazzianum* is known to flower at an age varying from 2–5 years or more. The clear effect of population stress in reducing the proportion of flowering at high population densities suggests a competitive effect during the previous years' vegetative growth. Samples taken after 2 years' growth revealed a hierarchy of biomass per plant at the higher populations, particularly in the roots. Individual plants had already gained random competitive ascendancy from an early stage of growth. The suppression of flowering observed in this trial and field observations suggest that a threshold root size (Weiner 1988) is required in *H. mantegazzianum* before reproduction is initiated as in *H. sphondylium* (Sheppard 1991). Small plants have been observed to remain vegetative for many years in relatively unfavourable (heavily shaded) habitats. This self-regulatory mechanism prevents overcrowding of flowering stems and allows a phasing of flower production from plants of different ages over several years, thereby improving dispersal strategy and at the same time adding to the difficulties of control. Inflorescence number and size, axillary branching and umbel components were all reduced at higher population densities, in common with the reproductive response to density in many plants (Weiner 1988). Total fruitlet production varies greatly from plant to plant, with location and with environmental conditions (Caffrey 1994).

Flower numbers up to c. 80,000 recorded here equate to a potential fruitlet production of 160,000. Fruitlet maturation is only a fraction of this (Tkatchenko 1989) and mature fruitlet counts ranged from 50,000-60,000 on large plants to 10,000-30,000 on average plants. The large numbers of viable seeds shed by *H. mantegazzianum* plants even when under environmental or management stress clearly makes prevention of flowering and fruiting a first priority in any control programme (Tiley and Philp 1992; Pyšek *et al.* 1995). The role of seed dormancy has yet to be evaluated.

Acknowledgments

The assistance of staff in Environmental Sciences Department, SAC Auchincruive and in BIOSS Scotland during the preparation of this paper is acknowledged. Support for these studies was received from the Scottish Office Agriculture Environment and Fisheries Department. Fig. 3 and parts of Tables 2, 4 and 5 are reproduced from the Journal of Ecology, Volume 84 with permission from Blackwell Science Ltd, Oxford.

References

Anonymous. 1970. Leading article. Lancet, 2: 32.

Caffrey, J.M. 1994. Spread and management of *Heracleum mantegazzianum* (Giant Hogweed) along Irish river corridors. In: L.C. de Waal, L.C. Child, P.M. Wade and J.H. Brock (eds.), Ecology and Management of Invasive Riverside Plants, pp. 67-76, Wiley and Sons, Chichester.

Caffrey, J.M. Unpublished information.

Camm, E, Buck, H.W.L. and Mitchell, J.C. 1976. Phyto photodermatitis from *Heracleum mantegazzianum*. Contact Dermatitis, 2: 68-72.

Clegg, L.M. and Grace, J. 1974. The distribution of *Heracleum mantegazzianum* (Somm. and Levier) near Edinburgh. Transactions of the Botanical Society of Edinburgh, 42: 223-229.

Dodd, F.S., Waal, L.C. de., Wade, P.M. and Tiley, G.E.D. 1994. Control and management of *Heracleum mantegazzianum* (Giant Hogweed). In: L.C. de Waal, L.C. Child, P.M. Wade and J.H. Brock (eds.), Ecology and Management of Invasive Riverside Plants, pp. 111-126, Wiley and Sons, Chichester.

Drever, J.C. and Hunter, J.A.A. 1970. Giant hogweed dermatitis. Scottish Medical Journal, 15: 315-319.

Gibson, S.W., Tiley, G.E.D. and Philp, B. 1995. The impact of invasive weeds on riverine flora in Scotland: a preliminary report. Report to Scottish Natural Heritage, Edinburgh.

Gray, E.G. and Noble, M. 1965. *Sclerotinia* disease. Scottish Agriculture, 44: 265-267.

Lee, T.D. 1988. Patterns of fruit and seed production. In: Lovett Doust, J. and Lovett Doust, L. (eds.), Plant Reproductive Ecology: Patterns and Strategies, pp. 179-202.

Lundström, H. 1989. New experiences of the fight against the Giant Hogweed, *Heracleum mantegazzianum.* Weeds and Weed Control, 30th Swedish Crop Protection Conference, 2: 51-58.

Lundström, H. and Darby, E. 1994. The *Heracleum mantegazzianum* (giant hogweed) problem in Sweden: Suggestions for its management and control. In: L.C. de Waal, L.C. Child, P.M. Wade and J.H. Brock (eds.), Ecology and Management of Invasive Riverside Plants, pp. 93-100, Wiley and Sons, Chichester.

Ochsmann, J. 1992. Riesen-Bärenklau (*Heracleum* spec.) in Deutschland. Morphologie, Ökologie, Verbreitung and systematische Einordnung. Diplomerbeit vorgelegt, Göttingen University.

Pyšek, P. 1991. *Heracleum mantegazzianum* in the Czech Republic – the dynamics of spreading from the historical perspective. Folia Geobotanica et Phytotaxonomica, 26: 439-454.

Pyšek, P., Ku-era, T., Puntieri, J. and Mandák, B. 1995. Regeneration in *Heracleum mantegazzianum* – response to removal of vegetative and generative parts. Preslia, Praha 67: 161-171.

Shumova, E.M. 1973. A study of morphogenesis of the hogweeds. Izvestiya Timiryazevskoi Selskokhozyaistvennoi Akademii, 1: 52-61 (in Russian).

Sheppard, A.W. 1991. Biological Flora of the British Isles: *Heracleum sphondylium L.* Journal of Ecology, 79: 235-258.

Stewart, F. and Grace, J. 1984. An experimental study of hybridization between *Heracleum mantegazzianum* Somm. and Lev. and *Heracleum sphondylium* L. spp. *sphondylium* (Umbelliferae). Watsonia, 15: 73-83.

Stewart, F. 1979. Hybridization between *Heracleum mantegazzianum* Somm. and Lev. and *Heracleum sphondylium* L. (Umbelliferae) in the British Isles. Ph.D. Thesis, University of Edinburgh.

Tiley, G.E.D., Dodd, Felicité S. and Wade, P.M. 1996. Biological Flora of the British Isles: *Heracleum mantegazzianum* Sommier and Levier. Journal of Ecology, 84: 297-319.

Tiley, G.E.D., and Philp B. 1992. Strategy for the control of giant hogweed *Heracleum mantegazzianum* on the River Ayr in Scotland. Vegetation Management in Forestry, Amenity and Conservation Areas. Aspects of Applied Biology, 29: 463-466.

Tkatchenko, K.G. 1989. Peculiarities and seed productivity in some *Heracleum* species grown in the Leningrad area. Plant Resources, 25: 52-61. (in Russian).

Weiner, J. 1988. The influence of competition on plant reproduction. In: Lovett Doust, J. and Lovett Doust, L. (eds.), Plant Reproductive Ecology: Patterns and Strategies, pp. 228-245.

Williamson, J.A. and Forbes, J.C. 1982. Giant hogweed, *Heracleum mantegazzianum*, its spread and control with glyphosate in amenity areas. In: Proceedings of the British Crop Protection Conference – Weeds, 1982, pp. 968-972.

HERACLEUM MANTEGAZZIANUM: A PROBLEM FOR RIVER MANAGERS IN THE REPUBLIC OF IRELAND AND THE UNITED KINGDOM

M. Wade[1], E. J. Darby[2], A.D. Courtney[3] and J.M. Caffrey[4]
[1]*International Centre of Landscape Ecology, Department of Geography, Loughborough University, LE11 3TU UK.*; [2]*Loughborough Ecologists, 226 Charnwood Road, Loughborough, LE12 9NR UK*; [3]*Department of Applied Plant Sciences, The Queen's University of Belfast, Newforge Lane, Belfast BT9 5PX, UK*; [4]*Central Fisheries Board, Balnagowan, Mobhi Boreen, Glasnevin, Dublin 9, Ireland*

Abstract

Heracleum mantegazzianum has been recognised for some time as a potentially serious invasive weed, particularly because of its phytophototoxic sap. This aspect of the plant is reviewed and the spread of the species at the national and local level is investigated. At the national level, *H. mantegazzianum* is still in an expansive mode in both the United Kingdom and the Republic of Ireland and an up to date distribution map is provided for Ireland. At the local level, a survey was undertaken in Northern Ireland in 1993 to re-examine known sites for the species and to search for new sites. The 96 original sites had declined to 43 (55%) but 29 new sites were reported, representing a net decline of 25%. The 72 sites existing in 1993 totalled 5.3 ha, 90% of which derived from nine sites of more than 1,000 m^2. The complexity of the spread of the species is considered in relation to management strategies for the future.

Introduction

"Mary and Matilda Blunt were itching and in pain under wet compresses ... Not really causes for a national panic. Yet at that time, it was not possible to say how many blistered children there may have been in England." (Holman, 1974).

Since its introduction in the 19th century, *H. mantegazzianum* has spread throughout the British Isles. Its preferred habitat has been the banks of water courses but it has quickly colonised waste places, railway embankments and roadsides. Apart from its propensity to suppress native flora, impact amenity areas and present problems for flood defence engineers, *H. mantegazzianum* can inflict a phytophotodermic reaction on those who come into physical contact with it. With the plant expanding its range and an increase in water-based activities it is inevitable that more people will come into contact with *H. mantegazzianum* in the future. Until recently, relevant authorities have shown little willingness to control *H. mantegazzianum* in order to reduce contact between the plant and people. The need for swift and co-ordinated action to reduce the number of contact accidents is examined and a strategy suggested for future control.

Plant Invasions: Studies from North America and Europe, pp. 139–151
edited by J.H. Brock, M. Wade, P. Pyšek and D. Green

H. mantegazzianum is a native of the Caucasus where it flourishes in clearings in upland forests. During the 19th century the plant was introduced to Western Europe and in the 100 years since its introduction to the British Isles the plant has spread, mainly *via* river corridors, creating a number of problems (Tiley and Philp 1994). The extent of this invasion is such that these problems have become serious, notably in parts of Scotland, south-east England and Ireland. Whilst the distribution of the species in England, Scotland and Wales is considered to have been kept up to date, the picture in Ireland is less clear (Tiley *et al.* 1996). Ireland has been described recently as the "Forgotten Island" with respect to botanical matters (Webb 1994) and attempts have been made to reassess the distribution of *H. mantegazzianum*, a species considered to be under-recorded from Ireland (Wyse Jackson 1995). This has involved not only identifying new sites but also determining whether or not the plant has maintained itself at a given site, the latter study being undertaken in Northern Ireland. Such reviews of the demography of the species are important in determining the type of management which is necessary to control the plant and minimise the problems it can cause.

Until recently, little or no attempt has been made to control the spread of *H. mantegazzianum* at a regional or national level. This probably reflects the fact that few people are aware of the health hazard posed by the plant and that, among those that are, few appreciate the rate at which the plant is spreading. Neither is it acknowledged that control will only be achieved through a comprehensive management strategy including all interested parties. The increased risk of litigation by people injured by the plant has prompted some Local Authorities to take action and implement control measures. However, such authorities are often unaware that the longer the problem is left unattended, the more difficult and more costly will be the ultimate control.

This paper considers the social problems caused by this plant and relates the growing concern about its harmful and problematic nature to its distribution in the United Kingdom and the Republic of Ireland. The implications of these findings are discussed in relation to the management of this species.

Incidence of phytophotodermatitis: the social dimension

A particular and unfortunate attribute of *H. mantegazzianum* is the ability of the sap to elicit a serious phytophototoxic response when it comes into contact with human skin in the presence of bright sunlight (Camm *et al.* 1976; Morton 1978; Rubow 1979; Lundström 1984). Early reports of dermatitis were from gardeners, e.g. in 1937 (Briggs 1979) and after some British Rail workers had suffered blistering when clearing a stand of the plant on a railway embankment. Two medical officers showed the blistering to be caused by sensitisation of the skin and the intensity of the ultraviolet light (Jones and Russell 1968) (wavelengths > 3,200Å, Briggs 1979). The sap remains phototoxic, even hours after the plant has been cut. More recently the National Rivers Authority (North-west region) recorded incidents of contact with the plant where blistering occurred without exposure to sunlight and where clothing appeared to offer little protection from the hazardous sap. Drever and Hunter (1970) described cases of *H. mantegazzianum* inducing phytophotodermatitis in children and adults from Edinburgh, Scotland. Symptoms appeared 24–48 hours after contact and

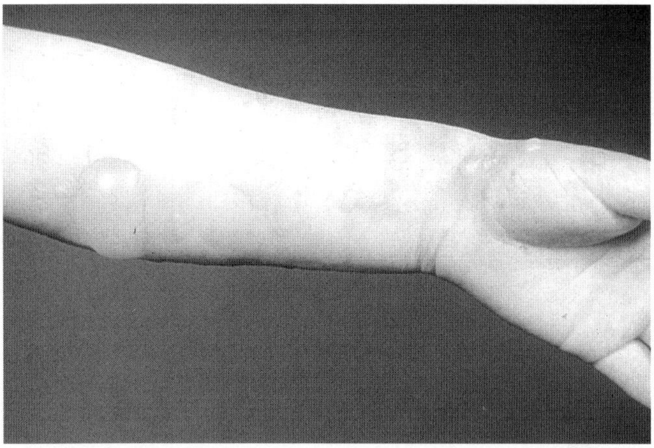

Plate 1. Example of blisters (bullae) on human skin after contact with *Heracleum mantegazzianum.*

involved a mild to severe erythematous reaction (red colouring of the skin) with or without painful blisters (bullae) (Plate 1). These were usually on the exposed parts of the body that came into contact with the plant. According to the medical literature, dense post-inflammatory hyper-pigmentation is visible within 3–5 days of contact and may persist for at least six years (Anon 1970). A recurrence of the condition may occur when the affected skin is again exposed to sunlight, even in the absence of further contact with the plant. An extensive medical literature has developed on the dermatological aspects of contact with *H. mantegazzianum* and is reviewed by Musajo and Rodighiero (1962); Pathak (1974) and Scott *et al.* (1976).

Numbers of recorded incidents of people suffering from contact with *H. mantegazzianum* are small. It is difficult to know if this is lack of recognition of the symptoms which are commonly dismissed as rashes of unknown origin. The difficulty of diagnosis is exacerbated by the time-lapse between contact with the plant and the appearance of rash and blisters. Where cases have been reported the symptoms are often serious enough for more than one medical opinion to be sought to determine the cause. For example, two boys holidaying on the estuary of the River Tweed at Berwick in Scotland reported severe itching on their arms after coming into contact with the plant. By the next day the itching had become intense and small blisters were in evidence. The local doctor diagnosed rash caused by contact with a hairy caterpillar. As the blisters became larger, "like large potatoes hanging off their elbows", the boys were taken to the hospital where a doctor recognised the symptoms and prescribed anti-histamine drugs. The boys subsequently recovered although dark blotchy patches persisted on their skin for many months.

In Ireland, *H. mantegazzianum* has been regarded for some time as a plant with "serious public health implications". St. Columcille's Hospital in Loughlinstown, Co. Dublin, which is close to large stands of *H. mantegazzianum*, regularly treats patients for *H. mantegazzianum*-induced burns (Wyse Jackson 1989). More recently, the media have focused on *H. mantegazzianum,* particularly when the plant has reached its maximum height, drawing attention to the hazards associated with it and its spread (e.g. Cusack 1993). Lucey (1994) describes stands of *H. mantegazzianum* which had colonised the roadside from the Newport River, Co. Tipperary, "as a hazard to

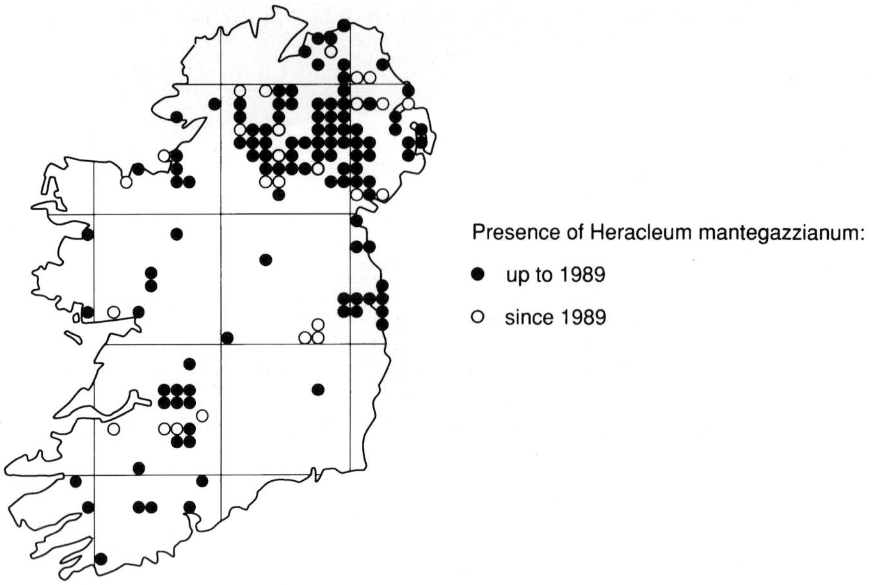

Fig. 1. Distribution of *Heracleum mantegazzianum* in Ireland, 1989 and post-1989.

pedestrians as well as motorists and in places restricting public access". In Northern Ireland, the Departments of Agriculture and Environment have responded to public concern through a campaign to increase public awareness of the potential hazards posed by this plant (Her Majesty's Stationery Office 1993).

Distribution and status in the Republic of Ireland

The first record of *H. mantegazzianum* in the British Isles was in Phoenix Park, Dublin, though the exact date is not known (Niven 1880). According to Colgan (1904) the plant was "naturalised" in the country at Blackrock Park, Dublin in 1902 and was "much grown in parks and shrubberies" at this time. A useful review of the distribution of *H. mantegazzianum* is provided by Scannell and Synnott (1987), a census which covered all of Ireland and records the species in 23 vice-counties. The first detailed account of *H. mantegazzianum* in Ireland, both the Republic and Northern Ireland, was by Wyse Jackson (1989). He found that the species had extended its distribution to a further six or seven vice-counties, a spread which had been primarily along river corridors (Caffrey 1994).

In order to update the distribution map for the species, all Botanical Society of the British Isles (BSBI) recorders for the Irish vice-counties were asked to return any records of the plant not included by Wyse Jackson (1989). These records were collated with those of Lucey (1994) and other sources of records (Harron 1986; Scannell and Synnott 1990; Hackney 1992; Wyse Jackson 1995). Lucey (1994) co-ordinated an extensive survey for *H. mantegazzianum* of some 1,800 sites on almost 400 rivers and streams in the southern counties of Ireland and added 24 new records, thus increasing the number of 10 × 10 km square records at that time from 15 to 20

Table 1. Heracleum mantegazzianum sites in N. Ireland, historical and current incidence.

County	Historical	Current incidence (1993)	
		Historical	New
Anthrim	8	0	0
Armagh	27	8	8
Down	22	18	1
Fermanagh	9	3	3
London-Derry	12	6	0
Tyrone	18	8	17
TOTAL	96	43	29

(Fig. 1). Along the Newport River, North Tipperary, where *H. mantegazzianum* has been present for more than 50 years (Praeger, 1939), the plant had achieved pest proportions. Of the five new 10 × 10 km square records, only one was considered as a new invasion, the River Galey, co. Limerick (Lucey 1994). It was first recorded here in 1993 but had not been observed on previous visits in 1982, 1986 and 1989. Collation of BSBI vice-county recorders' records combined with those found in other sources (Biological Records Centre, Institute of Terrestrial Ecology) gave a revised distribution for Ireland, with an additional ten 10 × 10 km squares (Fig. 1). The latter were mainly extensions of existing areas and included some new pre-1989 records. It would be instructive to return to these sites to check whether or not *H. mantegazzianum* has persisted and, if not, ascertain why it has disappeared.

While less specific data on the history of the plant at other sites are available, it is clear that the rate of spread has significantly increased along many river corridors in the past 30 years (Caffrey 1993a). Colonisation and spread have nearly always been in a downstream direction, reflecting the dependence of the plant upon seed dispersal by flowing water. Intensive examination of selected watercourses in recent years has shown no reduction in levels of infestation, but rather, in almost all cases, the populations have increased.

Implications for fisheries in the Republic of Ireland

On long sections of some of Ireland's more popular salmonid and coarse fishery rivers (e.g. Newport in Co. Tipperary, Mulcair in Co. Limerick, Bride in Co. Cork, and Dee in Co. Louth) dense bankside infestations of *H. mantegazzianum* have developed. These seriously restrict access to the water not only for anglers, a primary user group in a country renowned for the quality of its angling, but also boat owners, canoeists, swimmers, bird watchers, walkers and those working along river systems. Along these river corridors more or less continuous stands of this hazardous plant line the banks, including the water's edge. In most cases both banks are infested and safe access is precluded. The financial and social implications for angling clubs and local tourism are considerable.

In addition, and perhaps more seriously, the spread of *H. mantegazzianum* along rivers also threatens to destabilise the delicate equilibrium present in many productive salmonid rivers (Caffrey 1993b). Dense populations of this tall, leafy plant

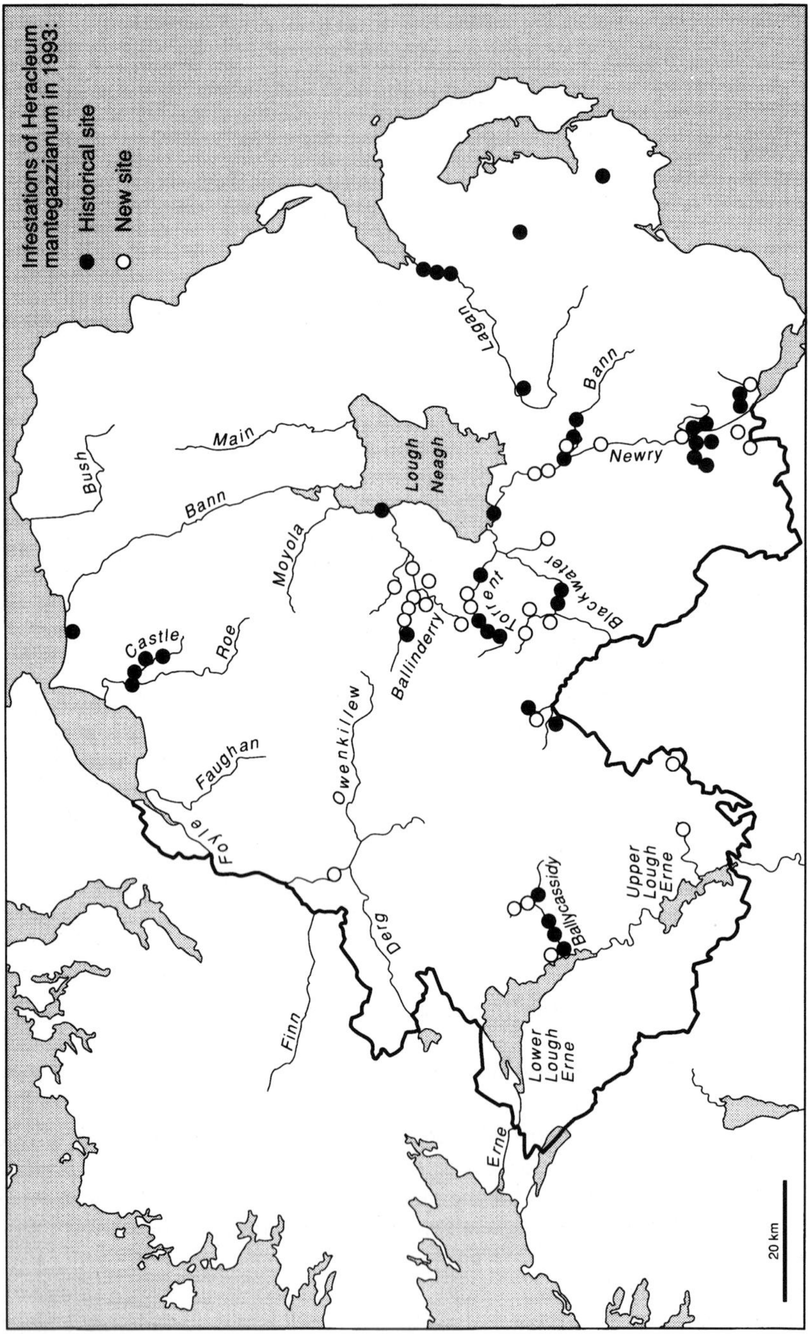

Fig. 2. Distribution of *Heracleum mantegazzianum* and river systems in Northern Ireland, 1993.

Fig. 3. The extent (estimated areas in m^2) and infestation levels (low, common, dominant) of *Heracleum mantegazzianum* in Northern Ireland.

rapidly suppress and ultimately exclude indigenous herbaceous plant species, the roots of which play an important role in river bank stabilisation. In winter *H. mantegazzianum* dies back and the soil is exposed to rain and high water flows. This can result in serious erosion problems with large amounts of soil being washed into the river. The relatively fine particulate suspension may settle in slow flowing sections of channel, where it will alter the substrate characteristics and, as a consequence, create favourable conditions for abundant aquatic plant growth (Caffrey 1990). Finer soil and silt particles clog gravel interstices and render them unsuitable for salmonid spawning. Research in Ireland (Gargan and Caffrey 1991; Caffrey 1992) and elsewhere (EIFAC 1974; Reiser and White 1988) has shown that salmon and trout eggs buried in gravel can only develop satisfactorily when a current of water is passing through the gravel. An oxygen deficit to the developing ova will result when silt causes a reduction in this intragravel percolation. Thus, by creating altered substrate characteristics or clogging spawning gravels with silt, indirectly *H. mantegazzianum* has the potential to seriously impact productive salmonid fisheries.

Distribution and status in Northern Ireland

Historical records of the type described previously provide only limited information on the area infested and the level of infestation. In order to provide baseline data on which to base advice and action on the potential spread of this species, a survey was undertaken of *H. mantegazzianum* in Northern Ireland in 1993 to establish more precisely the extent and levels of infestation. A list of all known stands of the species was circulated to surveyors, mainly staff of the Department of Agriculture Northern Ireland Watercourse Management and Applied Plant Sciences Research Division, and vice-county recorders of the BSBI, requesting reports on whether or not the stand was still present and on all new infestations within their area. Site details of habitat type (river or stream bank, field margin, roadside or other) and the area and level of infestation measured according to the scale: 0 = absent; 1 = low (occasional with one

Table 2. Incidence and extent (estimated areas in m²) of *Heracleum mantegazzianum* sites in Northern Ireland, 1993.

	Infestation level							
	Low		Common		Dominant		Total	
County	Area m²	Site No.	Area m²	Site No.	Area m²	Site No.	Area m²	Site No.
Armagh	17	(4)	722	(11)	20	(1)	759	(16)
Down	1,052	(4)	6,400	(7)	12,050	(8)	19,502	(19)
Fermanagh	28	(3)	12,035	(2)	150	(1)	12,213	(6)
London-Derry	0	(0)	122	(3)	60	(3)	182	(6)
Tyrone	23	(3)	395	(9)	20,110	(13)	20,528	(25)
TOTAL	1,120	(14)	19,674	(32)	32,390	(26)	53,184	(72)

to 5 plants visible within the infested area); 2 = common (6–20 plants visible within the area); 3 = dominant (plants too numerous to count) were also recorded.

Of the 96 sites reported from earlier records, 53 were no longer evident (Table 1). This represented a 55% decline which had occurred to a greater or lesser extent in all the counties and with no sites being known in County Antrim. New sites, 29 in total, failed to compensate fully for these losses, the net number of sites found in 1993 being 72 (Table 1). The majority of these sites were river or stream bank habitat (54 sites) but there were a number of instances where the plant had spread to adjacent field margins (9 sites) and roadsides (14 sites). The remaining eight sites were on waste ground, dumps and the banks of a lake.

The distribution of *H. mantegazzianum* sites (both persisting historical and new) (Fig. 2) illustrates a clear association of the main sites with specific river systems: the Ballycassidy, Ballinamallard, Trillick river system; the Ballinderry river system; the Blackwater river system; the River Bann; the Newry Canal and Clanrye river system; the River Lagan; and the River Roe and Castle river system. It is also important to note that other river systems have not as yet been invaded by this species, e.g. the River Foyle system and the River Main system. There was no obvious reason for the disappearance of *H. mantegazzianum* from certain sites. There may have been differences in management but these were not apparent.

The 72 sites recorded in 1993 comprised an estimated total area of 53,184 m² occurring mainly in the counties Down, Fermanagh and Tyrone (Table 2). Most of this total area was found in sites where *H. mantegazzianum* was dominant (32,390 m²) (Fig. 3). More interestingly, nine sites each with individual areas greater than 1,000 m² account for 90% of the total area infested (Fig. 4).

Local experience suggests that several of the larger infestations on the Bann at Gilford and at Trillick exhibit very invasive characteristics and are increasing in size. There are however many examples in the survey of sites which have remained at between 10–100 m² for more than 20 years (e.g. some sites on the rivers Ballinderry and Blackwater and all the sites on the River Roe). The Lagan infestations also remain quite contained and the relatively severe infestation on the upper reaches at Magheralin, whilst locally dominant and spreading onto adjacent farmland, does not appear to be serving as a focus for spread down the river system. The variation in

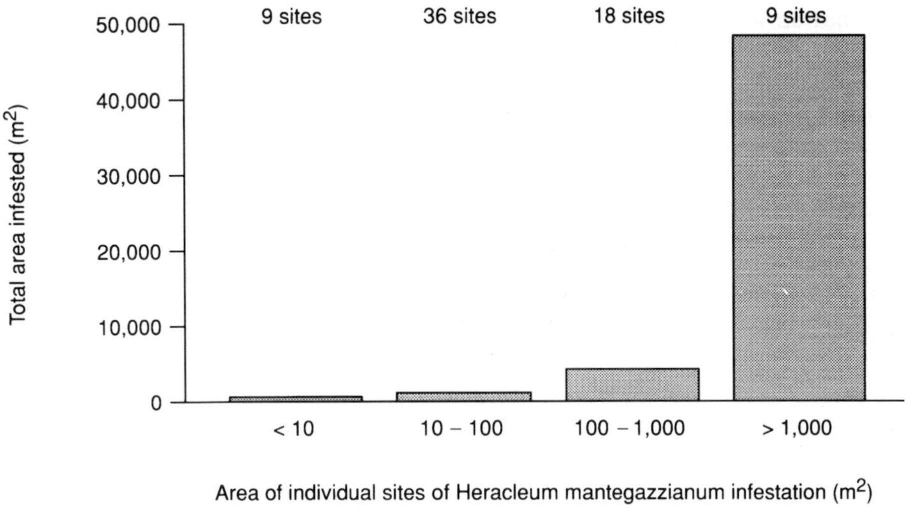

Fig. 4. Heracleum mantegazzianum incidence according to size of individual infestations (estimated areas in m^2).

Plate 2. Park bench in Hackney, London, with *Heracleum mantegazzianum* growing through slats.

demographic behaviour of *H. mantegazzianum* appears to show the larger infestations becoming locally more dominant but with several long established infestations failing to extend in an invasive fashion either locally or as a focus for new infestations. The particular site factors which restrict spread are unknown.

Some of the new records are of a size and intensity which indicates that they are long established, confirming the value of a systematic baseline survey.

Distribution and status in England, Scotland and Wales

Since its introduction into the British Isles in the nineteenth century (Niven 1880), *H. mantegazzianum* has spread rapidly to 119 of the 153 vice counties, with particular concentrations in south-eastern England and eastern Scotland (Tiley *et al.* 1996). In the last 50 years the distribution of the plant in England, Scotland and Wales has increased 40 fold, moving from streams and river valleys into the wider countryside (Tiley *et al.* 1996). Fowler *et al.* (1991) reported a doubling of recorded instances from 200 to 400 in the number of 10 × 10 km squares occupied between 1960 and 1980. Surveys carried out in the last 15 years in Scotland indicate an increase in the number of sites with consolidation of major infestations and fresh colonisation away from water courses (Neiland *et al.* 1987; Tiley and Philp 1992; Gibson *et al.* 1995). The present study does not attempt to update the distribution of *H. mantegazzianum* for England, Scotland and Wales beyond the latest distribution map given in Tiley *et al.* (1996).

Management strategies

The nature of the spread of *H. mantegazzianum* is more complex than first assumed. At the national level, the plant is still in an expansive mode, a pattern similar to that observed in other countries, for example the Czech Republic (Pyšek, 1991). This is well illustrated by its distribution history in Ireland (Fig. 1) and in England, Scotland and Wales (Tiley *et al.* 1996). Distribution and spread are mediated through river systems and, as there are still whole catchments in Ireland which are presently free of the plant (e.g. Fig. 2), it is not unreasonable to predict that further expansion is possible unless preventive action is taken.

At the local level, the pattern in Northern Ireland appears on the surface to be in reverse. The number of sites at which *H. mantegazzianum* can be found has decreased from 96 sites known prior to 1993 to 43 in 1993. Some 29 new sites were reported in 1993, representing a net decline of 25% though no data are available on changes in the total area infested. This decrease in sites suggests a transient nature of some *H. mantegazzianum* infestations or possibly the result of local control or management. The 1993 survey provides a baseline for future studies on the demography and management of the species in Northern Ireland. The overall decline in site records implies that mortality, natural or imposed, is exceeding establishment. A more specific understanding of the mechanisms of local spread and dispersal is desirable to formulate an effective control or eradication programme.

The control of *H. mantegazzianum* is technically straightforward (Dodd *et al.* 1994) but careful planning and implementation are essential. With the increased leisure time available to the public and the increased emphasis on outdoor and particularly water-based activities, it is predicted that more people will come into direct contact with *H. mantegazzianum* due to the overall increase in distribution of the species along water courses and its subsequent move into the wider countryside. The potential for expansion of *H mantegazzianum* appears to be primarily around, and downstream of existing areas of well established infestations. The control of these foci for expansion should be a first priority as it is with such large stands that the public are more likely to come into contact giving increased scope for litigation.

Such stands are expensive and time-consuming to control but prompt action is desirable to guard against further expansion into undisturbed habitats.

The central district of Scotland is particularly infested with *H. mantegazzianum* (Tiley *et al.* 1996), despite some successful attempts at local control (Neiland *et al.* 1987). Edinburgh City Council provides an excellent example of thinking through the measures that are available to managers to deal with the problem. The streams that run through the city and the closed cemeteries have been the principal dispersal pathways for *H. mantegazzianum.* Faced with the potential health risk posed by the presence of large areas of *H. mantegazzianum* in a crowded city, Edinburgh City Council has invoked the Public Health (Scotland) Act 1897 Part II, Section 16 (1). This defines a nuisance as "Any premises or part thereof of such a construction or in such a state as to be a nuisance or injurious or dangerous to health" and the City Council has issued enforcement notices requiring land owners to deal with *H. mantegazzianum* on their land (Steven Walker pers. comm. 1995). This policy has been pursued over a period of several years and so far it has not been necessary to go to court. Whilst being prepared to resort to law, the City Council pursues an active policy of plotting the location of *H. mantegazzianum*, pointing out the dangers (and the potential prosecution) to landowners and offering advice. Co-operation with the adjoining Lothian region has resulted in the production of a leaflet aimed at improving public awareness and an exchange of experience in control methodologies. *H. mantegazzianum* is now well known to the Edinburgh public who are apparently keen to co-operate in the control programme operated by the City Council.

This systematic and committed campaign is the only effective way to control *H. mantegazzianum.* There is a similar challenge for managers of river catchments for whom the responsibility is the same but perhaps more exacting. Managers need to mobilise public opinion and persuade landowners, other agencies and authorities to accept their involvement in what inevitably will be a long process. In 1995 the North West region of the NRA spent £30,000 on spraying the banks of the Rivers Crowle and Bollin in Manchester, UK, with glyphosate (David Holland pers. comm. 1995). The NRA carried out an extensive publicity campaign and received a good response from landowners. NRA (Southern region) also carried out a pilot spraying programme using glyphosate during 1995 on the Boldings Brook in West Sussex, UK (P. Griffiths pers. comm. 1995). Both NRA schemes can only be regarded as the beginning. Currently, individual authorities, such as the NRA, are carrying out localised control or trials. What is needed is a catchment scale co-ordinated control programme with follow-up monitoring. The future guarantee of a controlled area or catchment will also depend on adjoining catchments being treated in the same systematic and rigorous fashion. Future monitoring of the success of the initial spraying will need to be followed by several years of further control and monitoring given the potential seed bank (Tiley *et al.* 1996).

A good example of the need for a combined approach currently exists in the Lea valley, east London. *H. mantegazzianum* is densely distributed along the banks of the river and in the last ten years has invaded the adjacent land and now poses a threat to the local residents. The plant grows close to Hackney recreation ground, adjacent to the footpath and, in places, through the slats of the park benches (Plate 2). This represents a serious health hazard to the public and will result in injury if control measures are not implemented. Raising awareness is part of a strategy for control that must include long term co-ordinated action by all those responsible for public safety.

Any attempt to adopt half measures or failure to complete the course of treatment will result in greater effort and expense in the future.

Acknowledgements

The financial support of the Department of Agriculture Northern Ireland and in particular the staff of Water Course Management Division in the conduct of surveys in Northern Ireland and of Dr Trevor McCallion and Mr David Matthews (Biometrics Division) for the statistical analysis of these records is acknowledged. Thanks are also due to Dr John Faulkner, Department of the Environment Countryside and Wildlife Branch personnel, and other vice-county recorders of the Botanical Society of the British Isles who provided specific information on the incidence of *H. mantegazzianum* in Ireland. Gordon Tiley, Scottish Agricultural College, and Felicité Dodd kindly reviewed the paper.

References

Anon 1970. Leading article. Lancet 2: 32.

Briggs, M. 1979. Giant hogweed - a poisonous plant. BSBI News (21): 27-28.

Caffrey, J.M. 1990. The classification, ecology and dynamics of aquatic plant communities in some Irish rivers. Ph.D Thesis, University College, Dublin, 254 pp.

Caffrey, J.M. 1992. Salmon ova planting experiments in the River Deel catchment, North Mayo. North Western Regional Fisheries Board commissioned report, Central Fisheries Board, Dublin. 13 pp.

Caffrey, J.M. 1993a. Growth characteristics, invasive potential and control of Giant Hogweed in Ireland. Internal report, Central Fisheries Board, Glasnevin, Dublin, 14 pp.

Caffrey, J.M. 1993b. The spread and management of Giant Hogweed (*Heracleum mantegazzianum*) along Irish river corridors. Central Fisheries Board, Internal Report. Glasnevin, Dublin.

Caffrey, J.M. 1994. Spread and management of *Heracleum mantegazzianum* (Giant Hogweed) along Irish river corridors. In: de Waal, L.C., Child, L.E., Wade, P.M. and Brock, J.H. (eds) Ecology and management of invasive riverside plants. pp. 67-76. John Wiley, Chichester.

Camm, E., Buck, H.W.L. and Mitchell, J.C. 1976. Phytophotodermatitis from *Heracleum mantegazzianum*. Contact Dermatitis 2: 68-72.

Colgan, N. 1904. Flora of the County Dublin. Hodges and Figgis, Dublin.

Cusack, J. 1993. Warning on hogweed danger. The Irish Times, 22 June 1993, 1.

Dodd, F.S., de Waal, L.C., Wade, P.M. and Tiley, G.E.D. 1994. Control and management of *Heracleum mantegazzianum* (Giant Hogweed). In: de Waal, L.C., Child, L.E., Wade, P.M. and Brock, J.H. (eds) Ecology and management of invasive riverside plants. pp. 111-126. John Wiley, Chichester.

Drever, J.C. and Hunter, J.A.A. 1970. Giant hogweed dermatitis. Scottish Medical Journal 15: 315-319.

EIFAC 1974. Water quality criteria for European freshwater fish, Report on finely divided solids and inland fisheries. EIFAC Technical Paper (1) 21 pp.

Fowler, S.V., Holden, A.N.G. and Schroeder, D. 1991. The possibilities for classical biological control of weeds of industrial and amenity land in the U.K. using introduced insect herbivores or plant pathogens. Proceedings of the Brighton Crop Protection Conference, Weeds 1991, 3: 1173-1180.

Gargan, P. and Caffrey, J.M. 1991. Results of a preliminary survey of the Rivers Deel and Owenmore, Co. Mayo, with particular reference to peat siltation. North Western Regional Fisheries Board commissioned report, Central Fisheries Board, Dublin. 41 pp.

Gibson, S.W., Tiley, G.E.D. and Philp, B. 1995. The impact of invasive weeds on riverine flora in Scotland: a preliminary report. Unpublished report to Scottish Natural Heritage, Edinburgh.

Hackney, P. (ed.) 1992. Stewart and Corry's flora of the north-east of Ireland. Vascular plant and charophyte sections. 3rd edition. Institute of Irish Studies, Queen's University, Belfast.

Harron, J. 1986. Flora of Lough Neagh. Irish Naturalists' Journal Committee, Belfast and University of Ulster, Coleraine.

Her Majesty's Stationery Office. 1993. Problem plants: Giant Hogweed. Countryside management publication. HMSO, London.

Holman, F. 1974. The escape of the Giant Hogweed. Grasshopper Books, London.

Jones, J.G. and Russell, D.G. 1968. Giant hogweed dermatitis. Practitioner, 200: 704.

Lucey, J. 1994. Records of the Giant Hogweed, *Heracleum mantegazzianum* Sommier and Levier, along southern Irish rivers and streams with a revised distribution map for the region. Bulletin Irish Biogeographical Society, 17: 2-6.

Lundström, H. 1984. Giant hogweed, *Heracleum mantegazzianum*. A threat to the Swedish countryside. Weeds and Weed Control, 25th Swedish Conference, Uppsala, 1: 191-200.

Morton, J.K. 1978. Distribution of giant cow parsnip (*Heracleum mantegazzianum*) in Canada. Canadian Field Naturalist 89: 183-184.

Musajo, L. and Rodighiero, G. 1962. The skin photosensitizing furocoumarins. Experimentia 18: 153-161.

Neiland, M.R.M., Proctor, J. and Sexton, R. 1987. Giant hogweed (*Heracleum mantegazzianum* Somm. & Levier) by the River Allan and part of the River Forth. Forth Naturalist and Historian, 9: 51-56.

Niven, J.C. 1880. Cow parsnips. The Garden 17:33.

Pathak, M.A. 1974. Phytophotodermatitis. In: Fitzpatrick, T.B., Pathak, M.A., Harber, L.C., Seji, M. and Kukita, A. (eds) Sunlight and man. pp. 495-513. University of Tokyo Press, Tokyo.

Praeger, 1939. A further contribution to the flora of Ireland. Proceedings Royal Irish Academy 45B: 231-254.

Pyšek, P. 1991. *Heracleum mantegazzianum* in the Czech Republic - the dynamics of spreading from the historical perspective. Folia Geobotanica et Phytotaxonomica, 26: 439-454.

Reiser, D.W. and White, R.G. 1988. Effects of two sediment size-class on survival of Steelhead and Chinook salmon eggs. North American Journal of Fisheries Management 8: 432-437.

Rubow, T. 1979. Bekaempelse af Kaempe-bjornelko (*Heracleum mantegazzianum*). Statens Planteavlsforsog Meddelelser 81: 1-4.

Scannel, M.J.P. and Synnott, D.M. 1987. Census catalogue of the flora of Ireland. Ed. 2. The Stationery Office, Dublin.

Scannel, M.J.P. and Synnott, D.M. 1990. Records for the census catalogue of the flora of Ireland in the herbarium. National Botanic Gardens, Glasnevin. Occasional Papers 5: 1-42.

Scott, B.R., Pathak, M.A. and Mohn, G.R. 1976. Molecular and genetic basis of furocoumarin reactions. Mutation Research 39: 29-74.

Tiley, G.E.D., Dodd, F.S. and Wade, P.M. 1996. Biological Flora of the British Isles: *Heracleum mantegazzianum* Sommier and Levier. Journal of Ecology 84: 297-319.

Tiley, G.E.D. and Philp, B. 1992. Strategy for the control of giant hogweed (*Heracleum mantegazzianum*) on the River Ayr in Scotland. Aspects of Applied Biology, 29: 463-466.

Tiley, G.E.D. and Philp, B. 1994. *Heracleum mantegazzianum* (Giant Hogweed) and its control in Scotland. In: Waal, L.C. de., Child, L.E., Wade, P.M. and Brock, J.H. (eds.) *Ecology and Management of Invasive Riverside Plants.* pp. 101-109. John Wiley, Chichester.

Webb, D.A. 1994. The Forgotten Island. BSBI News (65): 24-25.

Wyse Jackson, M. 1989. Observations on the Irish distribution of a plant with serious health implications: giant hogweed (*Heracleum mantegazzianum* Sommier and Levier). Bulletin of the Irish Biogeographical Society, 12: 94-112.

Wyse Jackson, M. 1995. Annotated records for rare, critical or under-recorded vascular plant taxa from Ireland. Irish Naturalists' Journal, 25:44-57.

Hackney, P. (ed.) 1992. Stewart and Corry's Flora of the north-east of Ireland. Vascular plant and Charophyte sections. 3rd edition. Institute of Irish Studies, Queen's University, Belfast.
Harron, J. 1986. Flora of Lough Neagh. Irish Naturalists' Journal Committee, Belfast and University of Ulster, Coleraine.
Her Majesty's Stationery Office. 1991. Problem plants. Giant Hogweed. Countryside management [illegible]. HMSO, London.
[illegible]

DISTRIBUTION AND FREQUENCY OF *BIDENS FRONDOSA* VAR. *ANOMALA* IN THE INDUSTRIAL AREA OF THE RUHRGEBIET (GERMANY)

Peter Keil
Clausthaler Str. 11, 45145 Essen, Germany

Abstract

Bidens frondosa L., native in North America, is an invasive alien plant in Europe. Two varieties of *B. frondosa* L. are known in Germany. This study deals with the variety *B. frondosa* var. *anomala* Port. ex Fern. In autumn 1994 the riparian areas of the river Ruhr and parts of the Rhein-Herne-Canal (North-Rhine-Westfalia, Germany) were examined for the occurrence of *Bidens frondosa* var. *anomala*. The results showed that the "anomal" variety of *Bidens frondosa* L. proved to be more frequent than the "normal" variety. The relationship between the abnormal awns and the local distribution of the anomal variety is hard to correlate to any certain dispersal manner.

Introduction

Bidens frondosa L. is an annual Composite, with a native distribution in North America from Canada throughout the USA to Mexico.

In Europe, *B. frondosa* is an invasive alien plant. The first record of this taxa dates from 1777 (Lhotská 1966). In Central Europe the naturalization began at the end of the 19th century. Since then *Bidens frondosa* has been distributed throughout Europe (Meusel and Jäger 1992). The main pathways of the spreading are the great rivers, for example the Rhine (Germany), Elbe (Germany), Oder (Eastern Europe), Thames (Great Britain), Loire (France) and very intensivly along the canal-systems for inland shipping (Wagenitz 1964-1979). The distribution in Germany is shown in Fig. 1.

Bidens frondosa reproduces itself only from seeds and grows up higher than 20 dm (average 10 dm). Big individuals are able to produce more than 500 flower heads with more than 10,000 seeds (average are 50–100 flower heads, see Lhotská 1968).

Although many varieties of *Bidens frondosa* are known, most of them are variabilities, which appear at the same time at the same locality (ecology adaptions), so none of them can be used for taxonomic purpose (Sherff 1937). One variety with a constant characteristic is *B. frondosa* var. *anomala* Port. ex Fern., which differs from the variety *B. frondosa* var. *frondosa* in the upwardly barbed awns (Fig. 2 and 3).

This anomaly presents interesting aspects concerning seed dispersal, distribution and frequency compared to the normal variety.

In this respect the riparian areas of the lower river Ruhr and the Rhine-Herne-

Plant Invasions: Studies from North America and Europe, pp. 153–159
edited by J.H. Brock, M. Wade, P. Pyšek and D. Green

Fig. 1. Distribution of *Bidens frondosa* L. in Germany (Zentralstelle Floristische Kartierung – Datenbank Phanerogamen Unkorrigierter Datenstand Juni 1995).

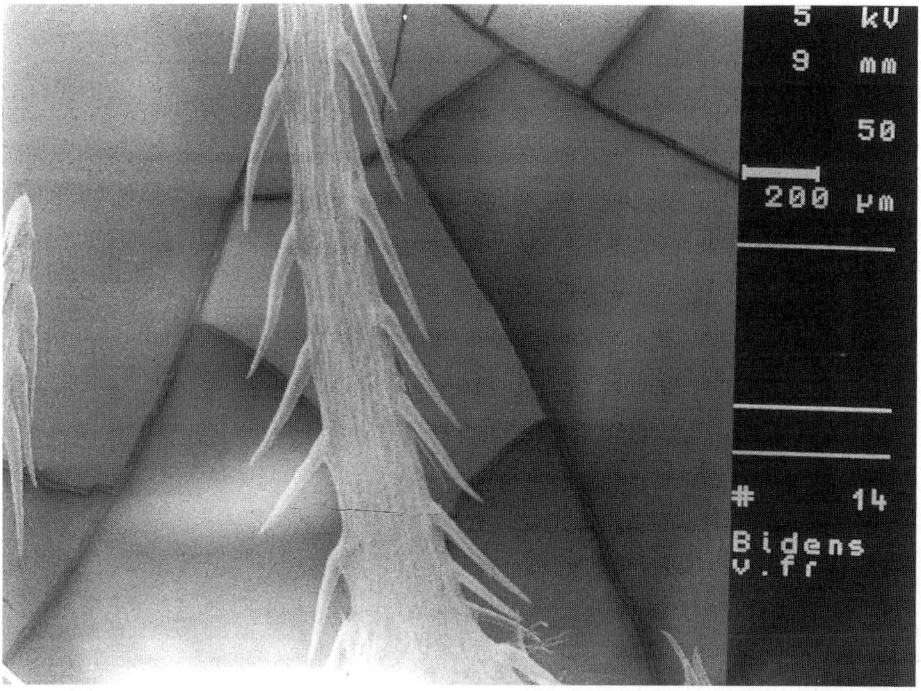

Fig. 2. Bidens frondosa var. *frondosa* with downwardly barbed awns.

Canal (both in western Ruhrgebiet, North-Rhine-Westfalia, Germany) were examined for the occurence of *B. frondosa* var. *anomala.* The main question was, what is the distribution and the frequency of *B. frondosa* var. *anomala* compared to the normal variety? If there is any difference, can we detect a disparity in how it is spread across the landscape?

The first records of *B. frondosa* var. *anomala* in the examined area date from Köstler in 1986 (Düll and Kutzelnigg 1987). In the whole of Germany only a few records are known from Westfalia and Berlin (Scholz 1955), Berlin (Sukopp and Markstein 1989), Hamburg, and in Baden-Württemberg (Ludwig 1994).

In neighbouring countries to Germany the occurrence of *B. frondosa* var. *anomala* has been documented since the 1960's from Poland (Trazcinska 1961), former Czechoslovakia (Lhotská 1964), The Netherlands (Oostroom & Reichgelt 1964) and Belgium (van Damme 1980).

Study area and methods

In autumn 1994 the riparian areas of the river Ruhr and the Rhine-Herne-Canal in Western Ruhrgebiet (North-Rhine-Westfalia, Germany) were examined. All individuals of *Bidens frondosa* which were found were exactly charted onto maps and counted (Table 1). The results of the mapping are documented by using the German topographic map system (MTB 1 : 25,000), which covers up an area of ca. 100 km^2 and can be subdivided into 16 quadrants (=¼ quadrants) of ca. 8 km^2 each (Fig. 4).

Table 1. Frequency of *Bidens frondosa* var. *frondosa* L. and *B. frondosa* var. *anomala* Port. ex Fern. In each quadrant, the more frequent variety is indicated by bold numbers.

¼-Quadrant-MTB	*Bidens frondosa* var. *frondosa*	*Bidens frondosa* var. *anomala*	Number of both varieties	variety *frondosa* (in %)	variety *anomala* (in %)
4407/44	0	2	2	0,00	**100,00**
4506/22	0	1	1	0,00	**100,00**
4506/23	0	13	13	0,00	**100,00**
4506/24	17	15	32	**53,13**	46,87
4506/32	38	102	140	27,14	**72,86**
4506/41	13	32	45	28,89	**71,11**
4506/42	6	10	16	37,50	**62,50**
4507/24	5	0	5	**100,00**	0,00
4507/31	5	2	7	**71,43**	28,57
4507/32	3	19	22	13,64	**86,36**
4507/34	17	41	58	29,31	**70,69**
4508/23	10	0	10	**100,00**	0,00
4508/32	10	10	20	50,00	50,00
4508/33	9	2	11	**81,82**	18,18
4508/34	51	36	87	**58,62**	41,38
4508/41	14	4	18	**77,78**	22,22
4508/42	26	18	44	**59,09**	40,91
4508/44	37	17	54	**68,52**	31,48
4607/12	23	23	46	50,00	50,00
4607/22	10	22	32	31,25	**68,75**
4607/23	4	57	61	6,56	**93,44**
4608/11	32	3	35	**91,43**	8,57
4608/12	46	27	73	**63,01**	36,99
Total	377	**455**	832	45,31	**54,69**
Number of the ¼-quadrants	20	**21**	23		

Results

Altogether, 832 individuals of *B. frondosa* were recorded and counted, and separated according to its recognizable varieties (Keil 1995). The results were: 377 individuals of *B. frondosa* var. *frondosa* and 455 individuals of *B. frondosa* var. *anomala*, that yields a proportion of about 45% : 55%. The variety "anomala" was more frequent than the "normal" variety (Table 1).

The local distribution of *B. frondosa* L. is heterogeneous. Both varieties are distributed in the riparian areas of the Ruhr. The distribution of those along the Rhine-Herne-Canal is more fragmentary, because there are many riparian constructions. *B. frondosa* var. *anomala* proved to be more frequent in the western while *B. frondosa* var. *frondosa* was more frequent in the eastern area (Fig. 4).

Most of the plants used habitats in the middle-water line of the Ruhr. But many individuals are also found all around the shore-lines, in riparian retention-walls and sometimes even outside of the river floodplain, for example in industrial areas, parks and along roadsides.

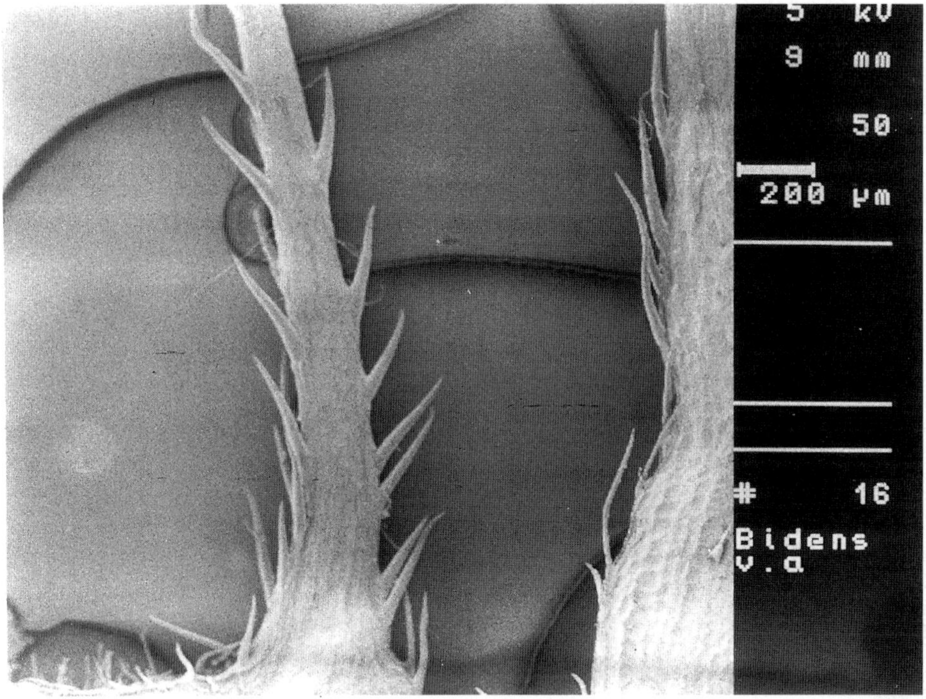

Fig. 3. Bidens frondosa var. *anomala* with upwardly barbed awns.

Discussion

In the study area *B. frondosa* var. *anomala* proved to be more frequent than *B. frondosa* var. *frondosa*.

This result agrees with observations from areas of the rivers Rhine and Main in Western Germany (Ludwig 1994) and the river Havel in Berlin (Scholz 1955).

The phytosociological relationship of *B. frondosa* var. *anomala* concerning the plant communities of the riparian areas is versatile. The main distribution can be localized within the *Calystegetalia* Tx. 50 in *Calystegion sepium* Tx. 47 communities. *B. frondosa* var. *anomala* is also able to occur in any other community along the riparian areas, for example on soils of the floodplain, in constructed river retention walls, and even in banks reinforced with concrete or stoney embankments. In the higher growing plant communities with the dominant invasive alien plant *Impatiens glandulifera* Royle, it was possible to observe individuals of *B. frondosa* var. *anomala*, which are able to reach a height from 230 cm. That shows, that they are also successful in competition with other dominant species.

The relationship between the anomal awns and the local distribution of the anomal variety is hard to correlate to any certain dispersal manner. The local frequency of *B. frondosa* var. *anomala* in the western area maybe due to hemerochor seed spreading (man's activity), that is favoured because of the conglomeration of industrial areas and traffic junctions with several highways, waterways and the large harbour Duisburg-Ruhrort.

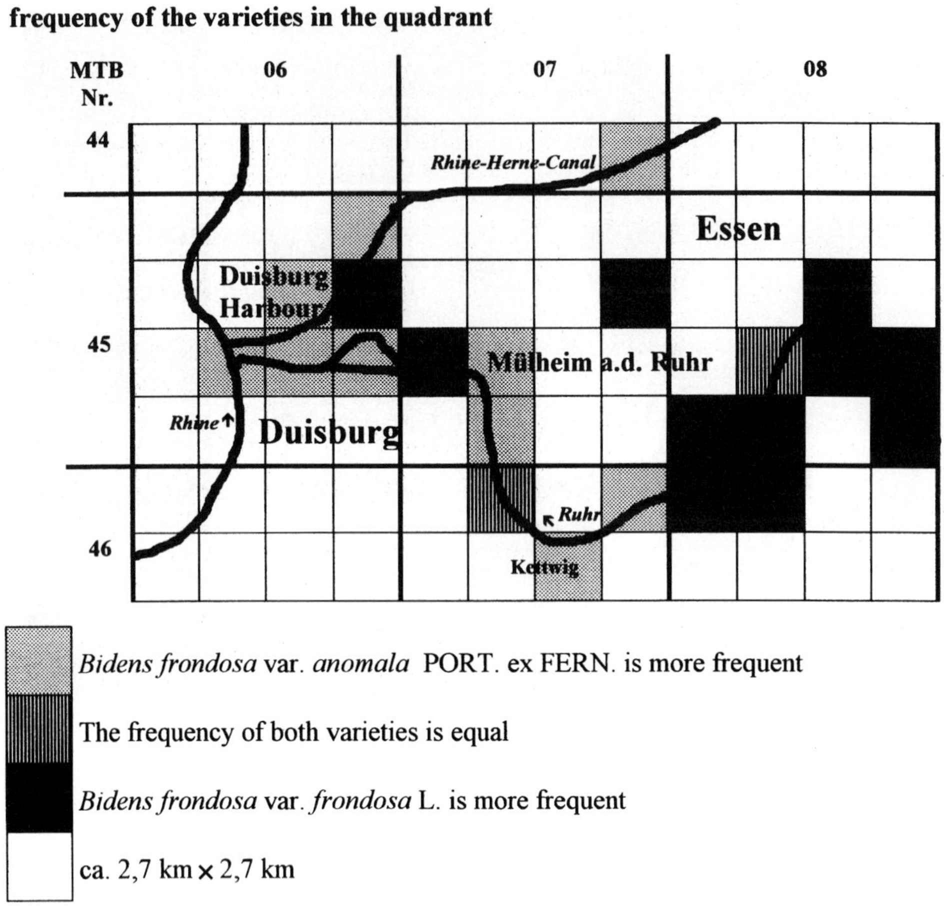

Fig. 4. Distribution of *Bidens frondosa* var. *anomala* Port. ex Fern. and *Bidens frondosa* var. *frondosa* in western Ruhrgebiet.

Acknowledgments

I thank J.H. Brock (Tempe, USA), H. Haeupler (Bochum, Germany), P. Pyšek (Průhonice, Czech Republic), T. Kordges (Hattingen, Germany) and C. Dübbert (Essen, Germany) for comments on the manuscript.

References

Düll, R. and Kutzelnigg, H. 1987. Punktkartenflora von Duisburg und Umgebung. 2. Aufl. – IDH-Verlag, Rheuth. 378 S.

Keil, P. 1995. Häufigkeit und Verbreitung von *Bidens frondosa* var. *anomala* PORTER ex FERNALD 1903 am Beispiel des unteren Ruhrtales im westlichen Ruhrgebiet. – Flor. Rundbr. (Göttingen) 29: 129-134.

Lhotská, M. 1964. *Bidens frondosa* L. var. *anomala* Port. ex Fern in Böhmen und Mähren. Preslia 37: 218-219 (in Czech).

Lhotská, M. 1966. Der älteste Fund der Art *Bidens frondosa* L. in Europa. – Folia Geobot. Phytotax. 1: 186-189.

Lhotská, M. 1968. Karpologie und Karpobiologie der tschechoslowakischen Vertreter der Gattung Bidens. – Rozpravy Československé Akademie Věd. ser. math.-nat., 1968/10: 1-85.

Ludwig, W. 1994. *Bidens frondosa* var. *anomala* in Hessen. Hess. Florst. Briefe 43(2): 17-19.

Meusel, H. and Jäger E.J., (eds.) 1992. Vergleichende Chorologie der zentraleuropäischen Flora 3. Jena, Stuttgart & New York.

Oostroom, S. J. and Reichgelt, T.J. 1964. Aanwisten voor de Nederlandse adventief-flora, 7. – Gorteria 2 (6): 65-67.

Scholz, H. 1955. Schwarzfrüchtiger Zweizahn (*Bidens frondosa* L.) ohne Klettfrüchte. – Natur u. Heimat (Münster) 15 (3): 83-85.

Sherff, E.E. 1937. The genus *Bidens*. Part 1. – Field Museum of Natural History, Botanical Series, Publication 388. Chicago. 1-346.

Sukopp, H. & Markstein, B. 1989. Vegetation der Berliner Havel – Bestandsveränderungen 1962 – 1987. Landschaftsentwicklung und Umweltforschung 64: 1-128.

Trazcinska, H. 1961. Studies on the distribution of synantropic plants. 1. *Bidens melanocarpus* Wieg. in Poland. Fragmenta Florist. et Geobot. 7: 161-168 (in Polish).

van Damme, C. 1980. Données concernant la répartition de *Bidens frondosa* L. var. *anomala* Port. ex Fern. (Asteraceae) en Belgique. – Dumortiera 17: 17-18.

Wagenitz, G. 1964-1979. *Compositae* I. In: Hegi, G. Illustrierte Flora von Mitteleuropa. Bd. VI/3. 3. Auflage. Berlin, Hamburg

INTRODUCTION AND NATURALIZATION OF *PRUNUS SEROTINA* IN CENTRAL EUROPE

Uwe Starfinger
Institut für Ökologie, TU Berlin, Schmidt-Ott-Str. 1, D-12165 Berlin, Germany

Abstract

Prunus serotina, a forest tree from eastern North America, was first introduced to Europe in the early 17th century and cultivated as an ornamental tree in parks and gardens. In the late nineteenth century foresters started to plant it in the forests. Contrary to their expectations it did not grow to merchantable timber size but rather built up dense shrub layers in oak-pine forests and invaded several types of woodland, especially on dry and poor sites. Now it grows spontaneously in many European countries. Today the species is considered a nuisance, because of its adverse effects on the natural rejuvenation of native forest trees and on the species richness of the ground layer. The major reasons for the naturalization success of *P. serotina* are:

1) The planting directly on the site of a possible naturalization, i. e. in the forests.
2) Its potential to germinate and grow under a very wide range of environmental conditions.
3) Its potential to produce fruit early in life and abundantly.
4) Its strategy of enduring as suppressed seedlings under a closed canopy.

It is likely that *P. serotina* will remain in the forests of Germany in the future, but with lower abundance than today. Because of its impact on forest ecosystems and on forestry further planting is not recommendable. Control, however, is difficult and expensive and should only be considered where clearly defined management or land use aims are threatened.

Introduction

From the extensive literature on "Biological Invasions" (e.g. Kornberg and Williamson 1986; Groves and Burdon 1986; Mooney and Drake 1986; Drake *et al.* 1989; Lohmeyer and Sukopp 1992; Pyšek *et al.* 1995; Cronk and Fuller 1995) it can be seen that introductions of organisms to areas where they were originally absent can have very different consequences. In the vast majority of cases the new species cannot survive in natural or near-natural vegetation. Only a very small percentage of the plant species introduced to Central Europe have naturalized at all. In some cases though, a new species can not only naturalize successfully, but also significantly change the whole ecosystem. In Central Europe the impact of exotic species on ecosystems is normally rather small, especially in natural and semi-natural communities such as forests.

One plant species that did spread in some parts of the continent is the American black cherry (*Prunus serotina* Ehrh.). In some parts of Germany and other European countries, foresters and people interested in nature conservancy have become increasingly concerned with it because it can become so abundant that it seems to threaten plant species of the forest floor and to interfere with the natural regeneration of forest

Plant Invasions: Studies from North America and Europe, pp. 161–171
edited by J.H. Brock, M. Wade, P. Pyšek and D. Green

trees. Studies on the invasion process of *P. serotina* and its effects have been published by Dutch researchers (e.g. Tweel and Eijsackers 1987) and by Starfinger (1990, 1991) in Germany. A short survey on the species in Europe was recently given by Scholz and Scholz (1995). In their worldwide invading species list Cronk and Fuller (1995) give *P. serotina* a rating of “4“ on an invasiveness scale of five parts, that is, a species “invading important natural or seminatural habitats“.

The objectives of this paper are to compare coexisting vegetation and demographic structure of *P. serotina* in plant communities in Pennsylvania, USA, and in Berlin, Germany, and to show:

a) the immigration and spread of *P. serotina* in the forests of central Europe,

b) the historical and ecological preconditions for this spread and

c) the consequences for flora and vegetation, for forestry, and other uses of the forest.

These results will allow us to evaluate the likelihood of further spread and to decide whether *P. serotina* is a “forest pest” or not. Recommendations will be given concerning possible further plantings and the necessity to control the species.

The species

P. serotina belongs to the family *Rosaceae*, subfamily *Prunoideae*. Within the genus *Prunus* it is usually grouped with others in the subgenus *Padus*. From other species in this subgenus it is distinguished by the persistent calyx, leaves in the inflorescence and toothed leave margins. The flowers on erect or spreading racemes appear in late spring, later than those of the European relative *P. padus*, hence the name (*serotina* = late flowering). The fruits are approximately 10 mm in diameter, ripe in late September or later, almost black when ripe and edible. The bark can be used as an ingredient of cough medicine. The plant itself can grow to a tall tree of 35 m in height and close to 1 m in diameter where the climate and soils permit, but it often remains a shrub or small tree. It can live up to about 250 years.

Prunus serotina in its natural range

Geographic distribution

P. serotina grows from Nova Scotia west to Minnesota, south to central Florida and Texas and in the mountains of Central America from Mexico to Guatemala. A distribution map was produced by Fowells (1965). It occurs from little above sea level in the coastal plains to the summits of the southern Appalachians at 1800–2000 m above sea level (McVaugh 1951). Specimens from the southern and the southwestern parts of this area have been described as different varieties, e.g. var. *eximia* Little 1953 or even as a different species, e.g. *P. capuli* Cav. ex Spreng. (=*P. serotina* ssp. *capuli* McVaugh 1951). In the main part of its range, the U.S. east of the Missisippi, however, only the typical form, *P. serotina* ssp. *serotina* occurs. Across this range *P. serotina* grows under a wide range of climatic conditions, with average yearly precipitation between 500 mm in the southwest and nearly 2000 mm in the

Appalachians. Average temperatures range from –15°C to 13°C in January and from 17°C to 28°C in July.

Sites and vegetation

The best developed stands on the Allegheny plateau grow on deep rich and moist soils. In other parts of its range such as the mid-western and southern states of the U.S., however, it occurs on poorer and dryer sands.

The forest types and associated tree species vary as much as the climatic conditions. The central part of the range coincides with the eastern deciduous forest (Braun 1950), but in the north, black cherry also grows in the transition zone to boreal forests. Near the Atlantic Ocean it grows in pine forests and in the west in oak savannas near the border of the prairies.

In a phytosociological field survey of the deciduous forests of eastern North America, Okuda (1994) lists *P. serotina* for most of the 24 tree communities he describes. It does not form a plant community by itself but reaches high frequencies in beech (*Fagus*) and maple (*Acer*) dominated communities in the northern part of the area of the deciduous forests and in the Appalachians as well as in oak-hickory (*Quereus-Carya*) forests in the piedmont. It is rare in southern forests and does not occur in boreal forests.

In many of these vegetation types *P. serotina* is either a minor component of the canopy or it does not reach canopy at all and remains in the shrub layer. Only at higher elevations in the southern Appalachians and especially on the Allegheny Plateau in Pennsylvania, New York, West Virginia, and Ohio it is of major commercial value. Here it occurs with minor frequency but as large trees in relicts of old-growth forests and can dominate a successional stage that developed from natural rejuvenation after clear-cuts in the late 18th and early 19th century. In this area, in the Allegheny National Forest in western Pennsylvania, Starfinger (1990, 1991) has conducted a phytosociological and demographical analysis recording the numbers of trees and of seedlings in four size classes and taking diameter measures of the trees. Small seedlings of *P. serotina* were regularly present under old conspecific trees, but larger seedlings and saplings were found only where larger trees were absent, that is where an opening in the canopy had been created. Cluster analysis of size classes showed the spatio-temporal distribution of the small seedlings to be different from the larger ones. This indicates a behavior known for a number of species of forest trees (Silvertown 1982): having short-lived seed these trees do not form seed banks (or only short-lived ones) in the floor like many other plants do, but instead build up a layer of regeneration underneath the canopy that can be called seedling bank. These seedlings can survive for a long time under adverse conditions (i.e. poor light) virtually without height growth. Only when the light conditions change, like after wind-throw or clear-cut of the canopy, these seedlings will start growing to eventually form the next generation of canopy trees. This behavior has been termed "Oskar Syndrome" by Silvertown (1982) after the character in the novel "*Die Blechtrommel*" by G. Grass, who stops growing at the age of three. The advantage of such an "Oskar strategy" lies in the fact that a tree species does not need to conquer a site after it becomes favorable for tree growth, but instead preempts the space and can thus be more successful than its competitors. So one might expect *P. serotina* to be a very successful species of old-growth forests. Its requirement for high light intensities in the sapling stage, however, limits its success in ecosystems that contain shade tolerant tree species such as beech (*Fagus grandifolia*) or some maples (*Acer* spp.). Consequently, *P. serotina* reaches dominance in the canopy only for a limited time during succession (Hough 1960). In old-growth forests in eastern North America it occurs with low frequencies only (Lutz 1930; Hough and Forbes 1943).

Several authors have shown that *P. serotina* is most often favored by disturbance of the ecosystem – be it man-made or natural (e.g. McBride 1973; Mackey and Sivec 1973; Skeen 1976).

In southern Wisconsin, *P. serotina* invaded oak forests that originated from oak savannas after white settlers controlled the formerly frequent fires. In this new ecosystem *P. serotina* is a frequent constituent of the understory. It was the object of a series of studies (Auclair and Cottam 1971, 1973; McCune and Cottam 1985). The authors predict that *P. serotina* will remain an important member of these forests, but find it difficult to predict its exact future role in the vegetation.

Prunus serotina **in Europe**

History of introduction

P. serotina was among the first tree species that were brought from North America to Europe. The first time it was cultivated in Europe was by Robin near Paris in 1623 (Wein 1930) or 1629 (Goeze 1916). In the following decades it was introduced to several European countries. In Germany, it was first recorded in 1685 (Wein 1930). For the following two centuries it was grown mostly as an ornamental in parks and gardens, due to its attractive flowers and autumn coloring. Its possible use for forestry in Europe was first mentioned in the late 18th century by von Wangenheim and von Burgsdorf (Booth 1896). Both recommended it as a species capable of producing valuable timber on poor soils.

Under the impression of a shortage of wood in the late 19th century, forest research stations in Germany began to test exotic tree species for their capacity to produce timber quicker than the native trees. *P. serotina* was among the species recommended for testing. At first it seemed to grow very well and fast under a wide range of conditions, including poor and dry soils (Booth 1896; Schwerin 1906). Later it was reported that the growth form was usually poor on these poor soils (Schwappach 1907).

Nevertheless the species was planted in the following decades, because it showed other advantages. Because its litter has a low C/N ratio (about 14) and decomposes quickly, it was used to improve the humus form in coniferous plantations. In the Netherlands it was planted widely from 1920 to 1950 for soil improvement and for fire and storm protection (Eijsakkers and Oldenkamp 1976). For Germany no data have been published concerning the time and extent of plantings apart from a few forestry records, but it can be assumed that *P. serotina* was planted on a large scale in the 1930s (cf. Ernst 1965). In the 1950s it was still recommended for planting in Germany (Wendorff 1952). In the 1970s and 1980s it was still used for fire and storm protection in some parts of Germany, also some forest research stations are still testing different geographic proveniences for timber production in Germany, assuming genetic, not only ecological reasons for the failure to grow well here.

Present distribution

Today *P. serotina* is present in large parts of the continental European lowlands, reaching high frequency and abundance in Germany, especially the northern half of the country, the Netherlands, NE France, Denmark, Poland (Stypinski 1979), and parts of Austria. It was also reported from Northern Italy (Abba 1975), Hungary, Romania (Ritiu *et al.* 1987), the Czech Republic, and Southern England (Stace 1991), but is apparently not present in Russia and adjacent states (Czerepanov 1995), the Mediterranean and Scandinavia.

Distribution maps have been published for Germany (Haeupler and Schönfelder 1988; Benkert *et al.* 1996) and the Netherlands (Tweel and Eijsackers 1987). From these it can be seen that its range coincides roughly with the distribution of sandy soils in both countries.

The naturalization process

Soon after the cultivation of the species the first spontaneously growing individuals were reported. For Berlin and Brandenburg in eastern Germany, Kowarik (1992) has collected data concerning introduction and spontaneous spread of *P. serotina*: It was mentioned as being cultivated here for the first time in 1796, and was found growing spontaneously in 1825. Thus it started its invasion process after a time-lag of only 29 years as compared to 147 years, an average of 184 woody species.

The fruit of *P. serotina* are eaten by several species of birds and mammals. Turcek (1961) lists 60 bird species that eat the fruit. Although birds and mammals act as long-distance dispersers, the majority of the seeds remain in the close vicinity of the mother trees. Smith (1975) found 71% of the seeds of *P. serotina* in traps less than 25 m away from the source trees. Indirect evidence for the distance travelled by the seeds can be found in the distance between seedlings and fruiting trees in areas where *P. serotina* is not growing very densely. Starfinger (1990) and Schulte and Schulze (unpublished data, cited in Kowarik 1995) in different regions of Germany both found a maximum of 600 m between seedlings and fruiting trees, the latter found more than half of the seedlings closer than 200 m and another third less than 400 m from seed sources. Kowarik (1995) points out that the spread of the species is affected by features of the ecosystem and calculates a rate of spread of 22.5 m/year in an open agro-ecosystem with hedges but significantly less in pine forests. He also stresses the importance of thinnings in the pine plantations, that is large scale disturbance of the forest ecosystem, that favor the invasion process. Borrmann (1987) reported on the spread of the species in a forest area in Mecklenburg in northern Germany where it was present on 438 ha. In only 52 ha of these it had originally been planted between 1910 and the late 1950s. He ascribes this eight-fold increase of the area covered by it in less than 80 years to spontaneous spread and invasion. Often it is difficult or even impossible to trace back the history of individual stands of *P. serotina*. Several authors have doubted that dense stands are mostly a result of an "aggressive invasiveness" of the species but may rather be connected to past planting practice (Tweel and Eijsakkers 1987; Starfinger 1990; Seidling 1993; Kowarik 1995).

A demographic analysis of the *P. serotina* stands in the forests of Berlin (Starfinger 1990) has shown a similar pattern to the one found in NW-Pennsylvania. It shows the same Oskar-behavior and the same trend towards stands with few large individuals and little regeneration as the population in Pennsylvania. In Berlin, however, the Oskars are taller than in Pennsylvania, which can be ascribed to the lighter canopy made up mainly by oak and pine and consequently better growth conditions for the seedlings. A comparison of the population biology of *P. serotina* in Pennsylvania and in Berlin is difficult because the ecological conditions and the vegetation of the two ecosystems are very different. A major difference lies in the history of the stands: whereas the forests in Pennsylvania result from natural regeneration after cuttings about 100 years ago, most of the forest trees in Berlin were planted and management by foresters results in repeated disturbance, such as thinning. After World-War II, many of the forests in and around Berlin were cut, so that conditions for the invasion of species were unusually good. Another difference lies in the fact that the role of *P. serotina* as a successional species in Pennsylvania is due to the fact that succession leads to dominance of shade tolerant trees. But this is not the case in Berlin. Here the annual precipitation of about 600 mm/year does not allow for shade tolerant trees such as the European beech (*Fagus sylvatica*) to become dominant. Oaks and pine, that build up the canopy, cannot regenerate successfully in the shade. In spite of these differences the two demographic analyses show that the ecological behavior of *P. serotina* is principally the same in both ecosystems. The basic similarity indicates that the same trend is present in the Berlin population. So it can be concluded from the comparison that frequency and dominance of *P. serotina* in the forests of Berlin will decrease in the course of time, but that the species will remain in the vegetation. The presence of *P. serotina* in the shrub and tree layer leads to a decrease of species richness in the ground layer. Starfinger has evaluated 301 phytosociological relevés from forests in Berlin made by several authors with the Braun-Blanquet (1964) method (sources in Sachse *et al.* 1990) and found a correlation between the presence of *P. serotina* in the shrub and tree layers with species richness (Fig. 1). Where *P. serotina* is present in the upper layers of the vegetation with a Braun-Blanquet cover value greater than “r”, the number of ground layer species per relevé decreases with increasing dominance of *P. serotina*.

Control

In some European countries or parts thereof, *P. serotina* is regarded as undesirable because of its effects on species richness of the forests, on natural forest dynamics, timber production or other aspects of forestry. These include the establishment of new forest stands by means of tree planting or natural rejuvenation. The presence of a dense shrub layer of *P. serotina* limits successful regeneration of the native forest trees, such as oak (*Quercus* sp.), pine (*Pinus sylvestris*), or birch (*Betula pendula*) which are mostly shade intolerant.

The control of *P. serotina* on the other hand poses a problem in itself. Cutting usually results in the production of basal sprouts which are often more vigorous and grow faster than the original plant. In order to kill the plants either chemicals or heavy mechanical control have been applied, both of which have unwanted side-effects and need to be repeated in order to treat regeneration.

In the Netherlands the problems connected with *P. serotina* were reported as early as 1963 by Bakker (1963), and led the author to call it “bospest” (forest pest). This

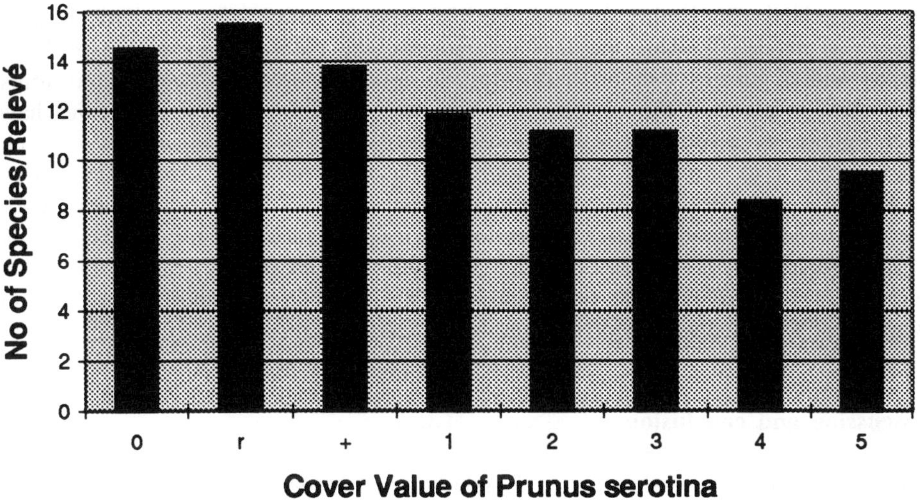

Fig. 1. Relation between the presence of *Prunus serotina* in the shrub and tree layers and the total number of ground layer species in phytosociological relevés from forests in Berlin. Cover values of Braun-Blanquet (1964): r: rare. 1 – 3 individuals, +: sparse. few individuals and little coverage, 1: plentiful, less than 5% coverage, 2: very numerous and less than 5 %, or 5 – 25 %, 3: 25 – 50 %, 4: 50 – 75 %, 5: 75 – 100 %. (after Starfinger in Sachse *et al.* 1990, changed)

resulted in more than 20 years of fighting the species with all means of control and in a number of research papers (Tol 1975; Eijsakkers and Oldenkamp 1976; Anon. 1984; Tweel and Eijsakkers 1987).

The first attempts at mechanical control soon proved to be unsatisfactory because of the sprouting response. Uprooting of younger plants was and is often done, partly with special tools designed for cherry control, because of the manpower needed it is very expensive (Mooyweer 1978). This, however, has negative effects on the soil, requires large amounts of manpower, machinery and money, and can lead to colonization by unwanted pioneer species.

In the 1950s–1960s, 2,4,5-T-herbicide was applied widely as a spray, until its use was prohibited because of its effects on the soil fauna (Eijsakkers 1974). Other herbicides such as amitrol (Jager and ten Kate 1975) and glyphosate (Roundup) (Jager 1977) were sprayed on the foliage partly with success, but because of side effects on non-target organisms this technique is no longer accepted.

Today the usual procedure to control *P. serotina* is a combined mechanical and chemical method. Trees are cut and glyphosate is applied to the stumps to prevent sprouting (Jager and Oosterbaan 1979), younger plants are pulled out. An alternative to the herbicide can be a biological control using the pathogenic fungus *Chondrostereum purpureum* that kills the plants (Scheepens and Zon 1982). The success of any of these control measures is uncertain, because *P. serotina* regenerates from numerous seeds that are germinable for at least four years (Eijsakkers and Ham 1984), and root fragments left in the soil. As Anon. (1984) points out, it is therefore necessary to repeat the control measures during the following years in order to clear a stand completely of *P. serotina*.

In Germany, various control methods have been applied with similar results as in the Netherlands. In Berlin forest authorities have a policy of naturalistic forestry (Krauß *et al.* 1990). This includes encouraging the natural rejuvenation of forest stands and a general ban on herbicides. In order to create a near-natural forest ecosystem in place of the existing post-war pine plantations forest authorities have been fighting *P. serotina* on a large scale since 1986. The general practice is to cut plants about 1 m above the ground and pull the roots and stub using either horses or a bulldozer. In Berlin, this costs approximately 1800,- DM/ha (horses) and 3000,- DM/ha (bulldozer), whereas the costs for a combined mechanical-chemical method (as described above) were estimated to be 300,- DM/ha (Spaeth *et al.* 1994). At present, *P. serotina* is much rarer than 10 years ago but seedlings are present, so, whether it will increase again remains to be seen.

Discussion and conclusion

The success of the naturalization of *P. serotina* in Europe leads to the question how the species differs from the many non-natives that are apparently unable to naturalize in a new area. Starfinger (1990) has discussed this extensively with the following results: *P. serotina* is a generalist in its germination requirements, is pollinated and distributed by a range of non-specialized insects and birds, produces abundant seed early in life and its seedlings can endure poor light conditions for a period of time. So it has a life history strategy that is favorable in central European oak-pine forests because there are no direct competitors with a similar strategy. It should, however, be taken into consideration that much of the present distribution of *P. serotina* is owed to planting directly in the forests and that its spread is favored by a history of strong disturbance of the recipient ecosystem. As in many other examples of invading species, the success of *P. serotina* can only be explained by a combination of characters of the species, of the recipient ecosystem, and circumstances of the introduction.

P. serotina is among the plant species that have not only naturalized in an area where it did not occur naturally, but have spread considerably and caused alterations in existing vegetation. Like many other plants in this group it is regarded as undesired, as an aggressive invader, and has even been called pest. Its effects on the native vegetation should be taken seriously and, as a consequence, further planting on a large scale is not recommended any longer. The judgment of it as an aggressive pest, on the other hand, seems exaggerated. From the demographic analysis reported in this paper it can be concluded that the species has conquered a niche in which no native species fits exactly and that it reaches dominance in the vegetation only during certain phases of the succession. Thus, an eradication program of *P. serotina* in Europe seems neither necessary nor feasible. Control may be reasonable in situations where its presence conflicts with specific management aims for a given forest plot, e.g. conservation of rare species or canopy regeneration. This, however, is true for other species (native or non-native) that can build up dense layers as well. *P. serotina* is as Joenje (1987) has put it "one of the species we must learn to live with".

Acknowledgments

The study was conducted at the Institute of Ecology of the Technical University Berlin. Thanks are due to my colleagues at the institute, above all to Herbert Sukopp for advice and encouragement. He and John Brock, Walter Seidling, Ingo Kowarik, and an anonymous reviewer helpfully commented on the manuscript. The research was partly supported by the local forest authority (Landesforstamt Berlin).

References

Abba, G. 1975. Some new and little known exotics in the Italian flora. Inf. Bot. Ital. 7: 362-364.
Anonymus. 1984. *Prunus serotina* van uitroeiing naar beheersing. Bosbouwvoorlichting 23: 77-81.
Auclair, A. N. and Cottam, G. 1971. Dynamics of Black Cherry (*Prunus serotina* Ehrh.) in Southern Wisconsin oak forests. Ecological Monogr. 41: 153-177.
Auclair, A. N. and Cottam, G. 1973. Multivariate analysis of radial growth of Black Cherry (*Prunus serotina* Ehrh.) in Southern Wisconsin oak forests. Am. Midl. Natur. 89: 408-425.
Bakker, J. 1963. De ontwikkelingsgeschiedenis van *Prunus serotina* in Nederland. Ned. Bosbouw Tijdschr. 35: 201-206.
Benkert, D., Fukarek, F. and Korsch, H. (eds.) 1996. Verbreitungsatlas der Farn- und Blütenpflanzen Ostdeutschlands. Gustav Fischer, Jena.
Booth, J. 1896. Die nordamerikanischen Holzarten und ihre Gegner. Zeitschr. f. Forst- u. Jagdwesen, p. 18.
Borrmann, K. 1987. Einbürgerung, Ausbreitung und Vorkommen der Späten Traubenkirsche (*Padus serotina* Borkh.) in der Oberförsterei Lütkenhagen (Kreis Neustrelitz). Bot. Rundbr. Bez. Neubrandenburg 19:13-18.
Braun, E. L. 1950. Deciduous forests of eastern North America. Hafner Publ. Co., New York and London.
Braun-Blanquet, J. 1964. Pflanzensoziologie. 3rd ed. Springer Verlag, Wien and New York.
Cronk, Q. C. B. and Fuller, J. L. 1995. Plant Invaders. The Threat to Natural Ecosystems. Chapman and Hall, London. 241 pp.
Czerepanov, S. K. 1995. Vascular Plants of Russia and Adjacent States (the former USSR). Cambridge Univ. Press, Cambridge, New York, Sidney.
Drake, J. A., Mooney, H.A., di Castri, F., Groves, R.H., Kruger, F.J., Rejmanek, M., Williamson, M. (eds.) 1989. Biological invasions. A global perspective. John Wiley and Sons, Chichester.
Eijsakkers, H. 1974. Mogelijke neveneffekten van bestrijdingsmiddelen tegen Prunus serotina. Ned. Bosbouw Tijdschr. 46: 55-62.
Eijsakkers, H. and Oldenkamp, L. 1976. Amerikaanse vogelkers, aanwaarding of beperking? Landbouwkundig Tijdschr. 88: 366-375.
Eijsakkers, H. and Ham, D. van den 1984. Kiemingspercentages van Amerikaanse vogelkers (*Prunus serotina* Ehrh.) onder laboratorium- en veldomstandigheden. Ned. Bosbouw Tijdschr. 56: 178-185.
Ernst, K. 1965. Späte Traubenkirsche und Traubenholunder. Berliner Naturschutz Blätter 26: 4-11.
Fowells, H. A. (1965): Silvics of forest trees of the United States. USDA Forest Service, Agriculture Handbook No 271.
Goeze, E. 1916. Liste der seit dem 16. Jahrhundert bis auf die Gegenwart in die Gärten und Parks Europas eingeführten Bäume und Sträucher. Mitt. Dtsch. Dendrol. Ges. 25: 129-201.
Groves, R. H. and Burdon, J. J. .1986. Ecology of biological invasions. Austr. Acad. of Science, Canberra.
Haeupler, H. and Schönfelder, P. (eds.) 1988. Atlas der Farn- und Blütenpflanzen der Bundesrepublik Deutschland. E. Ulmer, Stuttgart.
Hough, A. F. and Forbes, R. D. 1943. The ecology and silvics of forests in the high plateaus of Pennsylvania. Ecol. Monogr. 13: 299-320.
Hough, A. F. 1960. Silvical characteristics of Black Cherry, *Prunus serotina*. Station Paper No 139, NE Forest Exp. Sta. USDA Forest Service, Upper Darby, PA.
Jager, K. 1977. Glyfosaat, een nieuw middel voor de bestrijding van Amerikaanse vogelkers (*Prunus serotina*). Ned. Bosbouw Tijdschr. 49: 191-198.

Jager, K. and Oosterbaan, A. 1979. Bestrijding van Amerikaanse vogelkers (*Prunus serotina* Ehrh.) door middel van stam- of stobbebehandeling. Ned. Bosbouw Tijdschr. 51: 113-120.

Jager, K. and ten Kate, S. 1975. Nieuwe middelen voor de bestrijding van Amerikaanse vogelkers. Gewasbescherming 2: 25-31.

Joenje, W. 1987. Remarks on biological invasions. Proc. K. Ned. Akad. Wet. Ser. C 90, 15-18.

Kornberg, H. and Williamson, M. H. 1986. Quantitative aspects of the ecology of biological invasions. Phil. Trans. R. Soc. London B 314.

Kowarik, I. 1992. Einführung und Ausbreitung nichteinheimischer Gehölzarten in Berlin und Brandenburg. Verhandlungen des Botanischen Vereins von Berlin und Brandenburg (Berlin) Beih.3: 1-188.

Kowarik, I. 1995. Ausbreitung nichteinheimischer Gehölzarten als Problem für den Naturschutz? – In: Böcker, R.; Gebhardt, H.; Konold, W.; Schmidt-Fischer, S. (eds.), Gebietsfremde Pflanzenarten. Auswirkungen auf einheimische Arten, Lebensgemeinschaften und Biotope. Kontrollmöglichkeiten und Management. pp. 33-56. Ecomed, Landsberg.

Krauß, M., Loidl, H., Machatzi, B. and Wallacher, J. 1990. Vom Kulturwald zum Naturwald. Landschaftspflege konzept Grunewald. Berlin, Landesforstamt Berlin, 261 pp.

Lohmeyer, W. and Sukopp, H. 1992. Agriophyten in der Vegetation Mitteleuropas. Schriftenreihe für Vegetationskunde 25. Bonn-Bad Godesberg.

Lutz, H. J. 1930. The vegetation of Heart's Content, a virgin forest in northwestern Pennsylvania. Ecology 11: 1-29.

Mackey, H. E. and Sivec, N. 1973. The present composition of a former oak-chestnut forest in the Allegheny mountains of western Pennsylvania. Ecology 54: 915-919.

McBride, J. 1973. Natural replacement of disease killed elms. Amer. Midl. Natur. 90: 300-306.

McCune, B. and Cottam, G. 1985. The successional status of a southern Wisconsin oak woods. Ecology 66: 1270-1278.

McVaugh, R. 1951. Revision of the North American Black Cherries (*Prunus serotina* Ehrh. and relatives). Brittonia 7: 279-315.

Mooney, H. A. and Drake, J. A. (eds.) 1986. Ecology of biological invasions of North America and Hawaii. Springer Verlag, New York.

Mooyweer, A. M. 1978. De MORO Prunusrooier. Ned. Bosbouw Tijdschr. 50: 300.

Okuda, S. 1994. Deciduous Hardwood Forest Communities in Eastern North America, Including Some Conifer Forest and Shrub Communities. In: Miyawaki, A., Iwatsuki, K. and Grandtner, M.M. (eds.), Vegetation in Eastern North America. Vegetation System and Dynamics under Human Activity in the Eastern North American Cultural Region in Comparison with Japan. pp. 155-201. University of Tokyo Press, Tokyo

Pyšek, P., Prach, K., Rejmánek, M. and Wade, M. 1995. Plant Invasions – General Aspects and Special Problems. SPB Academic Publ. Amsterdam.

Ritiu, A., Nicolescu, L. and Nicolescu, N. 1987. Contributii la cunoasterea culturii malinului american (*Prunus serotina* Ehrh.) in nord-vestul tarii. (Contribution to the knowledge of American Black Cherry (*Prunus serotina*) in the north west of the country). Revista padurilor: Ser. Silvicultura si exploat area padurilor (Bucuresti) 102: 200-205.

Sachse, U. Starfinger, U. and Kowarik, I. 1990. Synanthropic Woody Species in the Urban Area of Berlin (West). – In: Sukopp, H., Hejny, S. and Kowarik, I. (eds.), Urban ecology. pp. 233-243. SPB Academic Publ., Den Haag.

Scheepens, P. C. and Zon, C. J. van 1982. Microbial Herbicides. In: Kurstak, E. (ed.): Microbial and Viral Pesticides. Marcel Dekker Inc., New York.

Schwappach 1907. Über die wichtigsten für deutsche Forsten geeigneten Laubholzarten. Mitt. Dtsch. Dendrol. Ges. 16: 126ff.

Schwerin, F. Graf von 1906. *Prunus serotina* Ehrhardt. Mitt. Dtsch. Dendrol. Ges. 15: 1-3.

Seidling, W. 1993. Zum Vorkommen von *Calamagrostis epigeios* und *Prunus serotina* in den Berliner Forsten. Verhandlungen des Botanischen Vereins von Berlin und Brandenburg (Berlin) 126: 113-148.

Scholz, H. and Scholz, I. 1995. *Prunus serotina*. In: Scholz, H. (ed.), Band IV, Teil 2B, Spermatophyta: Angiospermae: Dicotyledones 2 (3). Gustav Hegi Illustrierte Flora von Mitteleuropa. Ed. 2. Blackwell Sci. Publ., Oxford.

Silvertown, J. W. 1982. Introduction to plant population ecology. Longman, London.

Skeen, J. N. 1976. Regeneration and survival of woody species in a naturally created forest opening. Bull. Torrey Bot. Club 103: 259-265.

Smith, A. J. 1975. Invasion and ecesis of bird-disseminated woody plants in a temperate forest sere. Ecology 56: 19-34.

Spaeth, I., Balder, H. and Kilz, E. 1994. Das Problem mit der Spätblühenden Traubenkirsche in den Berliner Forsten. Allg. Forst- u. Jagd-Z. 11: 234-236.

Stace, C. 1991. New Flora of the British Isles. Cambridge Univ. Press, Cambridge. New York, Port, Chester, Melbourne, Sydney.

Starfinger, U. 1990. Die Einbürgerung der Spätblühenden Traubenkirsche (*Prunus serotina* Ehrh.) in Mitteleuropa. Landschaftsentwicklung u. Umweltforschung 69: 1-119.

Starfinger, U. 1991: Population Biology of an Invading Tree Species – *Prunus serotina*. In: Seitz, A. and Loeschke, V. (eds.), Species Conservation: A Population Biology Approach. pp 171-184. A. Birkhäuser Verlag, Basel.

Stypinski, P. 1979. Stanowiska czeremchy amerykansiej (*Padus serotina* (Ehrh.) Borkh.) w lasach panstwowych Pojezierza Mazurskiego. Rocz. Sekc. Dendrol. Pol. Tow. Bot. 32: 191-204.

Tol, G. van 1975. De Amerikaanse vogelkers. Gewasbescherming 6: 79-83.

Turcek, F. J. 1961. Ökologische Beziehungen der Vögel und Gehölze. Bratislava.

Tweel, P. A. van den and Eijsackers, H. 1987. Black Cherry, a pioneer species or 'forest pest'. Proc. K. Ned. Akad. Wet. Ser. C 90: 59-66.

Wein, K. 1930. Die erste Einführung nordamerikanischer Gehölze in Europa. Mitt. Dtsch. Dendrol. Ges. 42: 137-163.

Wendorff, G. von 1952. Die *Prunus serotina* in Mitteleuropa. Eine waldbauliche Monographie. Diss. Hamburg.

MONITORING INVASIVE WEEDS AT LANDSCAPE LEVEL IN DENMARK

Ulla Vogt Andersen
COWI Consulting Engineers and Planners, Parallelvej 15, 2800 Lyngby, Denmark

Abstract

The occurrence of invasive aliens was recorded in a survey conducted in south west Denmark of natural and seminatural areas. All localities were classified according to habitat type and a comprehensive list of vascular plants was drawn up for each of 61 localities. More than 50% of all localities studied were found to have been invaded by alien invasive species. A significantly higher number of localities with invasive species was found among the dry habitat types, than among the wet habitat types. Heaths were the most invaded of the habitat types. *Prunus serotina* was found to be the most prevalent invasive weed in this region of Denmark. This type of survey provides an overview of the distribution of invasive species in the landscape and may be useful in the plans for management of invasive weeds and can be used in an early warning system predicting possible invasive weed candidates. It can also be applied in the control strategies for specific invasive plant species.

Introduction

A number of plant species occur as invasive alien weeds in natural and semi-natural habitats throughout northern and central Europe. Such species are generally defined as non-native species, having the ability to establish in natural or semi-natural habitats, where they can eliminate native vegetation through competition (Pyšek 1995).

Some temperate climate habitats have been found to be more susceptible to invasion than others, for instance habitats, with a high degree of disturbance (Dean *et al.* 1994; DeFerrari and Naiman 1994). Most aliens occupy artificial, temporary habitats such as ruderal sites, wastelands, urban areas, road verges, and railway tracks (Sykora 1990). Only few are able to invade and establish in natural plant communities (Sykora 1990; Pyšek 1995). Natural plant communities, such as forests, are therefore regarded as relatively resistant to invasion by exotic species, due to their high degree of stability. Danish coastal communities, however, appear to have a high percentage of aliens, compared to other natural and semi-natural vegetation types (Andersen 1995b). The aims of the present study are therefore to investigate which of the natural or semi-natural habitats are most vulnerable to invasion, and to study the relationship between invasibility, size and species diversity in invaded localities.

Patterns of alien weed spread indicate that many species have a long lag phase following introduction before they become invasive (Kowarik 1995). An overview of invasive species at a landscape level may prove useful in the prediction of which

Plant Invasions: Studies from North America and Europe, pp. 173–181
edited by J.H. Brock, M. Wade, P. Pyšek and D. Green

species may become the future invaders. Only few studies have concentrated on the problems of invasive alien plant species at a landscape level (DeFerrari and Naiman 1994; Cherrill *et al.* 1995). It is therefore important to study whether or not the invaded localities are aggregated around areas under high human impact such as cities, roads, and railways, and to investigate the distribution of some selected invasive species in the landscape.

As rivers and streams have played an important role in the spread of invasive aliens in many parts of the world (DeFerrari and Naiman 1994; Pyšek and Prach 1995), streams, rivers and associated meadows are included in this survey to determine if these plant communities influenced by freshwater are more likely to have a higher number of invasive species, than other Danish plant communities.

Study area

The study was conducted in the Southwest region of Denmark in a 50 km long and 3–6 km wide corridor from Gredstedbro to the border to Germany. The region consists of material deposited by the ice during the second last glacial period. The soils are generally sandy podzols which in some places are overlain with organic material and most of the land lies between –1 to +40m above sea level. The region is crossed by five major rivers as well as a number of smaller streams. The land-use is primarily arable agriculture and includes rotation grasslands used for dairy cattle. Some coniferous forests have been planted in the last 80–100 years to prevent sand drift. The climate is very windy with the prevalent winds from the west over the North Sea.

Most of the communities studied are semi-natural, but a few small fragments of ancient woodlands consisting of *Quercus robur* persist. Some places with high ground water table are covered with mesotrophic fens or meadows, but this depends on the management practice. In places with moist, acidic soils, oligotrophic moors have developed, whereas on dry podzols, typical heaths dominated by *Calluna vulgaris*, *Erica teralix*, *Empetrum nirgum* or *Deschampsia flexuosa* have developed.

Methods

The present study is based on the field investigations carried out as a part of an environmental impact assessment (EIA) for a planned motorway between Gredstedbro and the border to Germany. The field survey was carried out during the growth season of 1995. The localities were chosen on a basis of 1:25000 maps as well as aerial photographs.

All woods, plantations, fens, meadows, rivers, lakes, heaths and moors were inspected and a comprehensive list of vascular plant species was collected from 61 localities. The localities were divided in two groups: those with more than 50 and those with less than 50 species. In this region of Denmark undisturbed localities usually have few species, whereas a high number of species can be a sign of disturbance, eutrophication etc. All localities were classified according to their major habitat type. If, for instance, a meadow contained a small waterhole it was still classified as meadow. Only semi-natural and natural areas were included in this

Table 1. List of invasive plant species found in southwest Denmark and the number of locations where they appeared.

Species	Number of locations
Prunus serotina	14
Pinus mugo	13
Rosa rugosa	10
Aegopodium podagraria	7
Acer pseudoplatanus	4
Rubus laciniatus	3
Heracleum mantegazzianum	2
Solidago canadensis	2
Spiraea douglasii	2
Petasites hybridus	1

survey, but no distinction was made between semi-natural and true natural as most Danish localities to some extent are under human impact. The localities were divided in two groups according to their size i.e. smaller than 5 ha, and larger than 5 ha, as all lakes, heaths, moors, meadows and freshwater fens larger than 5 ha are protected by the nature protection law

All localities were recorded on a digital map using the GIS-system Intergraph Microstation. The presence/absence of invasive alien vascular plant species was recorded for each locality together with other information such as area, number of species, and conservations. All information was kept in a database and *via* a locality number linked to the map. In the statistical data processing likelihood-ratio chi-square tests were used. The plant nomenclature follows Hansen (1981) except for *Heracleum mantegazzianum* Somm. & Lev.

Results

A total of 10 invasive alien plant species was found in the study area (Table 1). The woody species such as *Prunus serotina*, *Pinus mugo*, and *Rosa rugosa* are the most widespread in this part of Denmark and more than 50% of the localities (i.e. 34 of 61) were found to contain invasive species (Fig. 1). The distribution of invaded and non-invaded localities seems to show a random pattern. Table 2 contains a list of all localities studied, as well as the number of species, and their content of invasive aliens. Table 3 presents the result of the statistical tests.

The distribution of invaded and non invaded localities with more than 50 species or less than 50 species, was compared. There was no significant difference between these two locality types indicating that the presence or absence of invasive weeds is independent of the diversity of vascular plant species in the locality.

The distribution of localities with invasive species compared with localities without invasive species was not significantly different between localities larger than 5 ha, and those smaller than 5 ha, indicating that the presence or absence of invasive aliens is independent on the size of the locality. It was also tested to determine whether or not the floristic diversity of the locality was independent of the size. This hypothesis was accepted as the difference between the large and the small areas in relation to diversity was not significant (χ^2 = 3.349, p = 0.067).

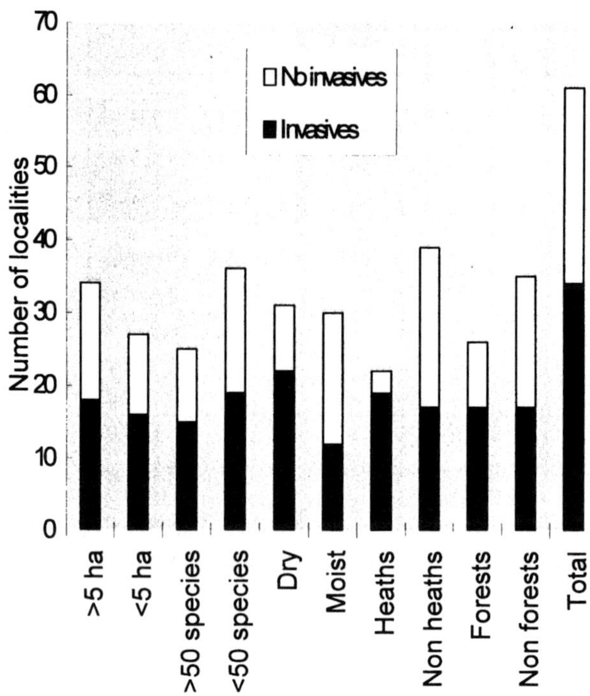

Fig. 1. Comparison of number of localities with and without invasive species. (For further explanation see caption to Table 3).

The occurrence of invasive species in heath localities compared to other localities was strongly significant (Table 3), indicating that the heaths are the most susceptible to invasion of the vegetation types studied. There was a significantly higher occurrence of invasive species in dry habitat types compared to those of wet and moist habitats. In contrast, there was no significant difference between the number of forested localities with invasive species and those without invasive species.

Discussion

Exotic species are believed only to occur in artificial habitats and only rarely to become a fixed element of the vegetation (Sykora 1990). The present study reveals, however, that invasion occurs in many semi-natural and in some natural communities. Forests usually provide very stable ecosystems and are therefore considered to be less susceptible to invasion. The forests of southwest Denmark, however, appeared to be neither more or less invaded than other habitat types, even if their environmental conditions remain unchanged for 80–120 years as in the case of coniferous plantations and for centuries in the "undisturbed" *Quercus robur* forests.

The random pattern of invaded and non-invaded communities in the landscape indicates that those localities which have been invaded are not aggregated around cities or distributed along major communication corridors, even if humans are the most important vectors for the spread of invasive aliens (Andersen 1995a). The

Table 2. List of 61 localities with locality number, habitat type (f.fen = freshwater fen; dec.for. = deciduous forest; con.plant. = coniferous plantation; mead. = meadow), area, total number of vascular plant species (= N), and invasive weeds present in the locality.

Location	Type	Area	N	Invasive species
1-5	dec.forest	<5 ha	29	–
2-1	mead/f.fen	<5 ha	84	*Rosa rugosa*
2-11	heath	<5 ha	23	–
3-1	heath/dune	>5 ha	85	–
3-2	heath/dune	>5 ha	69	*Pinus mugo* *Rosa rugosa*
3-3	meadow	>5 ha	100	*Petasites hybridus*
3-4	dec.forest	>5 ha	65	*Acer pseudoplatanus* *Aegopodium podagraria*
4-1	lake	<5 ha	23	–
4-2	dec.for/plant.	>5 ha	66	*Rubus laciniatus* *Rosa rugosa* *Acer pesudoplatanus* *Prunus serotina*
5-1	meadow	>5 ha	36	–
6-2	heath/plant.	>5 ha	80	*Prunus serotina* *Pinus mugo* *Rubus laciniatus*
6-4	dec. forest	<5 ha	25	–
6-7	con.plant.	>5 ha	42	*Prunus serotina*
6-8	heath/plant.	<5 ha	47	*Pinus mugo*
6-9	con. plant.	<5 ha	46	–
6-10	heath/plant.	<5 ha	24	*Rubus laciniatus* *Pinus mugo*
6-11	common	>5 ha	77	*Prunus serotina*
6-13	dec.forest	<5 ha	34	*Pinus mugo* *Prunus serotina*
7-1	heath	>5 ha	37	*Pinus mugo* *Prunus serotina*
7-2	fresh.fen	<5 ha	24	–
7-3	con.plant.	>5 ha	33	*Pinus mugo*
7-6	lake	<5 ha	26	*Rosa rugosa*
8-1	dec.forest	>5 ha	65	–
8-3	heath/f.fen	<5 ha	93	–
8-5	dec.for/lake	<5 ha	52	–
8-6	dec.for/lake	<5 ha	46	*Solidago canadensis* *Aegopodium podagraria* *Spiraea douglasii*
8-8	heath/plant.	>5 ha	77	*Rosa rugosa* *Pinus mugo* *Prunus serotina*
8-10	heath/dec.for.	<5 ha	54	*Spiraea douglasii* *Rosa rugosa* *Solidago canadensis*
9-5	heath	<5 ha	62	*Pinus mugo*
9-6	con.plant	>5 ha	73	–
9-8	heath/dune	>5 ha	22	–
9-9	heath/lake	<5 ha	42	*Rosa rugosa*
9-10	heath	<5 ha	25	*Rosa rugosa* *Prunus serotina* *Pinus mugo*

Table 2. Cont.

Location	Type	Area	N	Invasive species
9-11	dec.for./mead	>5 ha	27	–
9-13	heath	<5 ha	30	–
9-14	heath	<5 ha	34	*Prunus serotina*
9-16	heath	>5 ha	21	*Pinus mugo*
9-18	f.fen/mead.	>5 ha	53	–
11-1	f.fen	<5 ha	78	*Aegopodium podagraria*
11-2	plant./f.fen.	>5 ha	43	–
11-5	f.fen/mead.	>5 ha	47	–
12-1	con.plant.	>5 ha	65	–
14-2	lake/mead.	>5 ha	50	–
14-4	lake/mead.	<5 ha	33	–
14-6	meadow	>5 ha	24	–
15-3	river	>5 ha	29	*Heracleum mantegazzianum*
17-3	heath/plant.	>5 ha	63	*Prunus serotina* *Pinus mugo*
17-4	fresh.fen.	>5 ha	114	*Prunus serotina*
17-5	lake/plant.	<5 ha	90	*Rosa rugosa* *Acer pseudoplatanus* *Prunus serotina*
17-6	fresh.fen	>5 ha	94	–
17-7	river	>5 ha	24	–
17-8	dec.forest	<5 ha	30	*Acer pseudoplatanus* *Aegopodium podagraria*
17-11	heath/lake	<5 ha	82	*Heracleum mantegazzianum* *Rosa rugosa*
18-1	heath/plant.	>5 ha	26	*Pinus mugo*
18-5	heath/plant.	>5 ha	42	*Prunus serotina* *Aegopodium podagraria*
18-9	mead./lake	<5 ha	33	–
18-10	heath/dec.for.	<5 ha	31	*Prunus serotina*
18-6	fresh.fen	>5 ha	31	*Aegopodium podagraria*
18-7	fresh.fen	<5 ha	35	–
19-1	river	>5 ha	52	–
19-2	river	>5 ha	22	*Aegopodium podagraria*

Table 3. Results of the likelihood-ratio χ^2-test for invaded and non-invaded localities: Area: localities larger than 5 ha compared with those smaller than 5 ha. Diversity: localities with more than 50 species compared with those with less than 50 species. Type: dry habitat types compared with wet/moist habitat types. Heaths and forests compared with other vegetation types.

Test	χ^2	p
Area: >5 ha/<5 ha	0.244	0.621 ns
Diversity: <50 species/>50 species	0.313	0.576 ns
Type: dry localities/moist localities	6.025	0.015 *
Vegetation: heaths/non heaths	11.621	0.001 ***
Vegetation: forests/non forests	1.725	0.189 ns

dispersal of invasive aliens at a landscape level may therefore be achieved by means other than human activities. If disturbed, ruderal communities had been included in this study, the majority of invaded localities might have been more aggregated around the cities, as these community types are more prevalent in such areas.

The "top-ten" invasive species found in the present study differed from those which are generally assumed to be the most important invasive weeds in Denmark (Andersen 1995a). If the same type of study were to be carried out in the eastern part of Denmark, *Heracleum mantegazzianum*, *Solidago canadensis* and *Petasites hybridus* would probably have appeared as the most widespread among the invasive weed species. The reason for the abundance of *Prunus serotina*, *Pinus mugo* and *Rosa rugosa* is related to the sandy podzols in south west Denmark. The abundance of these species may also be related to the heath vegetation and the rather poor conditions in which this habitat type is found today. The major threats to heaths have for many years been cultivation and/or afforestation (Hansen 1964). The afforestation is mainly due to ceased grazing (Böcher and Jørgensen 1972). Eutrophication arising from precipitation and from fertilisation of adjacent areas has also influenced the vegetation.

It is generally assumed that invasion by aliens is encouraged by disturbance in plant communities (Dean *et al.* 1994). In the heaths of southwest Denmark the opposite seems to be the case: Invasion may be initiated by cessation of special management regimes such as grazing or peat cutting. Grazing with a low stocking rate may initiate the establishment of woody species in meadows and heaths (Buttenschön and Buttenschön 1982a,b), habitats that would otherwise be overgrown with tall herbs and grasses. During the grazing the woody species are kept down by the browsing, but when grazing ceases their cover increases. In contrast communities dominated by tall herbs and grasses are very resistant to natural colonisation and may therefore also be resistant to invasions by alien species (Silvertown *et al.* 1994; Smith and Rushton 1994). Plant communities may therefore be most vulnerable to invasions just after changes in the management regime, whereas a continuous disturbance often suppresses the invader.

The significantly higher number of invaded localities among the dry habitat types, is due to the high susceptibility to invasion of the heaths. This is different from the findings in other studies that show that wet and moist localities, especially riverbanks, generally containing a higher number of alien species than dry localities (DeFerrari and Naiman 1994). Although the rivers function as dispersal corridors for water dispersed species, there is no indication that invasive species are spread by watercourses in this part of Denmark. In accordance with the findings of DeFerrari and Naiman (1994), the present study reveals no relationship between the size of locality and invasibility.

Prunus serotina is the most important invasive weed in this region. This may be due to the past and present use of the species in hedges and windbreaks. Today the shrub is legally propagated in nurseries and planted in the open landscape, even though it is recognised as a potential suppresser of native species and vegetation. Further spread of *Prunus serotina* to new localities is easily achieved by birds. According to Rejmanek (1995) invasiveness of woody species, in both disturbed and "undisturbed" areas, is significantly associated with small seed mass and vertebrate dispersal special features possessed by *Prunus serotina.* The species is a widespread invasive weed in north Germany (Kowarik 1992).

Several authors have stated that a focus on the invaded ecosystem and the causal factors promoting invasion, rather than the individual invasive species, should be more efficient in preventing future problems (e.g. Hobbs and Humphries 1995). From my point of view both aspects are equally important, as some communities which might be regarded as safe from invasion may turn out to be invaded by species, which would never be invasive under other circumstances. Species, such as *Rubus laciniatus,* that are currently not abundant, but are present as invaders in some localities, should receive more attention.

The present study has provided a new way of overviewing the spread of invasive weeds at landscape level. When using GIS we have the possibility for combining vegetation data with other information such as land use, preservation regulations and ownership as described by Child and de Waal (this volume).

Acknowledgements

The author is very grateful to the colleagues at COWI for their participation in the collection and processing of field data. The study was financed by The Ministry of Traffic as a part of an environmental impact assessment for a planned motorway.

References

Andersen, U.V. 1995a. Comparison of dispersal strategies of alien and native species in the Danish flora. In: Pyšek, P; Prach, K.; Rejmanek, M. and Wade, M. (eds.). Plant invasions – General aspects and special problems. pp. 61-70. SPB Academic Publishing, Amsterdam.

Andersen, U.V. 1995b. Invasive aliens: a threat to the Danish coastal vegetation? In: Healey, M.G. and Doody, J.P. (eds.). Directions in European coastal management. pp. 335-344. Samara Publishing Limited, Cardigan.

Buttenschön, R.M. and Buttenschön, J. 1982a. Grazing experiments with cattle and sheep on nutrient poor, acidic grassland and heath. 1: Vegetation development. Natura Jutlandica 21 (1-3): 1-48.

Buttenschön, R.M. and Buttenschön, J. 1982b. Grazing experiments with cattle and sheep on nutrient poor, acidic grassland and heath. 2: Grazing impact. Natura Jutlandica 21 (2): 19-27.

Böcher, T.W. and Jørgensen, C.A. 1972. Jyske Dværgbuskheder. Eksperimentelle undersøgelser af forskellige kulturindgrebs indflydelse på vegetationen. Kongelige Danske Videnskabernes Selskab. Biologiske Skrifter 19(5): 55 pp.

Cherrill, A.J., McClean, C., Watson, P., Tucker, K., Rushton, S.P. and Sanderson, R. 1995. Predicting the distribution of plant species at the regional scale: a hierachial matrix model. Landscape Ecology 10: 197-207.

Dean, W.R.J., Milton, S.J., Ryan, P.G. and Moloney, C.L. 1994. The role of disturbance in the establishment of indigenous and alien plants at Inaccessible and Nightingale Islands in the South Atlantic Ocean. Vegetatio 113: 13-23.

DeFerrari, C. and Naiman, R.J. 1994. A multiscale assessment of the occurrence of exotic plants on the Olympic Peninsula, Washington. Journal of Vegetation Science 5: 247-258.

Hansen, K. 1964. Studies on the regeneration of heath vegetation after burning-off. Botanisk Tidsskrift 60: 1-41.

Hansen, K. (ed.) 1981. Dansk feltflora. Gyldendal, Copenhagen.

Hobbs, R.J. and Humphries, S.E. 1995. An integrated approach to the ecology and management of plant invasions. Conservation Biology 9: 761-770.

Kowarik, I. 1992. Einführung und Ausbreitung nichteinheimischer Gehölzarten in Berlin und Brandenburg. Verhandlungen des Botanishen Vereins von Berlin und Brandenburg. 1859 e.V.

Kowarik, I. 1995. Time lags in biological invasions with regard to the success and failure af alien species. In: Pyšek, P.; Prach, K.; Rejmanek, M. and Wade, M. (eds.). Plant invasions – General aspects and special problems. pp. 15-38. SPB Academic Publishing, Amsterdam.

Pyšek, P. 1995. On the terminology used in plant invasion studies. In: Pyšek, P; Prach, K.; Rejmanek, M. and Wade, M. (eds.). Plant invasions - General aspects and special problems. pp. 71-81. SPB Academic Publishing, Amsterdam.

Pyšek, P. and Prach, K. 1995. Invasion dynamics of *Impatiens glandulifera* a century of spreading reconstructed. Biological Conservation 74: 41-48.

Rejmanek, M. 1995. What makes a species invasive? In: Pyšek, P; Prach, K.; Rejmanek, M. and Wade, M. (eds.). Plant invasions - General aspects and special problems. pp. 3-13. SPB Academic Publishing, Amsterdam.

Silvertown, J., Lines, C.E.M. and Dale, P. 1994. Spatial competition between grasses - rates of mutual invasion between four species and the interaction with grazing. Journal of Ecology 82: 31-38.

Smith, R.S. and Rushton, S.P. 1994. The effects of grazing management on the vegetation of mesotrophic (meadow) grassland in Northern England. Journal of Applied Ecology 31: 13-24.

Sykora, K.V. 1990. History of the impact of man on the distribution of plant species. In: di Castri, F., Hansen, A.J. and Debussche, M. (eds.). Biological invasions in Europe and the Mediterranean Basin. pp. 37-50. Kluwer Academic Publishers, Dordrecht.

FIFTEEN YEARS OF CHANGES IN THE REPRESENTATION OF ALIEN SPECIES IN CZECH VILLAGE FLORA

Petr Pyšek and Bohumil Mandák
Institute of Botany, Academy of Sciences of the Czech Republic, CZ-252 43 Průhonice, Czech Republic; e-mail: pysek@ibot.cas.cz

Abstract

Floras of several villages in the Bohemian Karst Protected Area (Czech Republic) are compared over the time period of almost fifteen years and the trends in the performance of alien plant species are described. Floristic list from 1980-81 was compared with the recent situation in 1995. Alien species were divided into archaeophytes, consisting of species introduced before 1500, and neophytes, introduced after 1500. From 1982 to 1995, there was an increase in the total number of species recorded in the 10 villages studied by 26.3 % (from 407 in 1982 to 514 in 1995). The most remarkable increase (by 90 %) was recorded in neophytes. This group at present not only consists of higher number of species (95 in 1995 compared to 50 in 1982) but particular species are also more abundant than in the past. These trends may be explained by (i) an increase in tourism bringing more diaspores into the region, (ii) more plants available on the gardening market compared to the period 15 years ago, and (iii) boom of the building activity.

Introduction

The flora of human settlements harbours relatively high proportions of alien plant species (Sukopp and Werner 1983; Kowarik 1990; Reidl and Dettmar 1993). This is from the increased possibilities of dispersal and rather heterogeneous environment providing plant species with a variety of habitats (Gilbert 1989; Wittig 1991). Human-induced disturbances are also frequent in the city environment and although not unequivocally accepted, there is some agreement on the supporting role of disturbances for the establishment of alien species (Hobbs 1989; Rejmánek 1989; Kowarik 1995). Furthermore, high vegetation dynamics create a number of early successional stages in which type of environment increase the chances for alien species to establish (Rejmánek 1989).

Most of the floristic data available from human settlements in Europe concern towns and cities (Pyšek 1993). Such data in the majority of cases results of either a short-term record from a single vegetation period, or concern long-term research, in some cases covering more than a decade (e.g. Berlin, Germany – Kunick 1974, 1982; Kowarik 1990). In both cases, it is difficult to infer the changes in the flora composition in time; in the short-term studies because the records are unrepeated, in those long-term because the data are cumulative which makes it difficult to compare unless a very long time-interval is considered. However, historical data making it possible to draw quantitative conclusions are very rare (Klotz 1987; Kowarik 1990).

Plant Invasions: Studies from North America and Europe, pp. 183–190
edited by J.H. Brock, M. Wade, P. Pyšek and D. Green

The present paper concentrates on the representation of alien species in the flora of village settlements. With the villages, it is possible to make an up-to-date record of the flora present in a given time. The repetition of such a sampling enables analysis of changes in the floristic composition on a rather short time scale. The present study compares the floras of several central Bohemian villages over the time period of almost fifteen years and is aimed at describing the trends in the performance of alien plant species. Also, with the study area located within former Eastern block of Europe, the paper is concerned with the following question: Are the changes in life style which occurred in the Czech Republic as a consequence of political changes reflected in the features of human-associated flora?

Study area and methods

Field research was carried out in the Bohemian Karst Protected Area, central Bohemia, located approximately 30 km SW of Prague. The area has an annual temperature of 8.4°C, and an annual precipitation of 480 mm (50-years average, Beroun meteorological station). The following 10 villages were studied (with the number of inhabitants and number of inhabited houses given in the parentheses, data from 1980): Karlštejn (1082/249), Srbsko (535/144), Vysoký Újezd (337/101), Choteč (287/80), Bubovice (279/79), Chýnice (244/65), Lužce (132/33) Kuchař (116/34), Hostim (108/37), and Svatý Jan pod Skalou (78/20). The villages are located at the altitude of 225–432 m. A detailed phytosociological study on vegetation of the villages was carried out by Pyšek (1991, 1992).

In 1980–81, the floristic list including all species of higher plants was compiled for the territory of each village by the first author (Pyšek 1982, 1985). The margin houses at the village periphery were considered as the limits of the territory studied, and the species beyond these limits were not included in the list. In 1995, the floristic research was repeated using the same methods and criteria.

Total floristic lists (i.e. the lists of all species present in the set of villages studied) were compiled for each period and compared with respect to the representation of alien species. The species matching the following definition were considered as alien (introduced, exotic, adventive): a species which reached the area as a consequence of the activities of neolithic or post-neolithic man or of his domestic animals (Webb 1985; Pyšek 1995). The term is opposite to indigenous (native), i.e. those species which evolved in the area or which arrived there by one means or another before the beginning of the neolithic period or which arrived there since that time by a method entirely independent of human activity (Webb 1985). Alien species were further divided into two groups: (a) archaeophytes, consisting of species introduced before 1500, and (b) neophytes, introduced after 1500 (Holub and Jirásek 1967). The species status was determined using local floras and data bases (Frank and Klotz 1990, Ellenberg *et al.* 1991; Hejný and Slavík 1988-1992; Slavík 1995).

Neophytes were further analysed from the viewpoint of the area of origin, life form and life strategy based on Grime *et al.* (1988).

Table 1. Comparison of species richness according to the status between the two records (1982, 1995). Total species numbers recorded in the ten villages studied are given. The term Disappearances is used for those species present in 1982 but not found in 1995. Those not present in 1982 but recorded in 1995 are termed Newcomers.

	1982	1995	Disappearances	Newcomers
Indigenous	229	278	37	86
Archaeophytes	128	141	8	21
Neophytes	50	95	9	54
Total	407	514	54	161

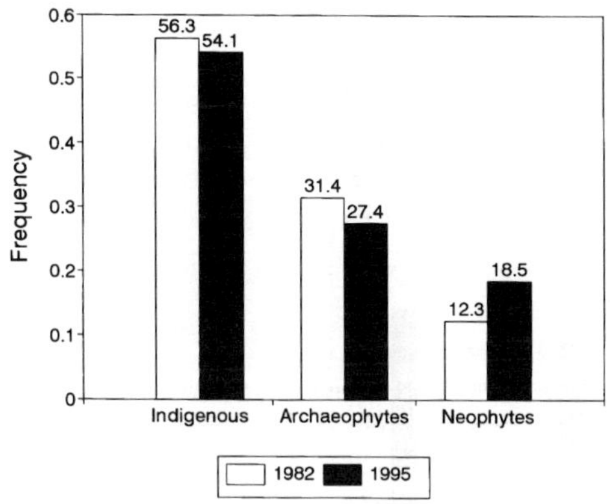

a

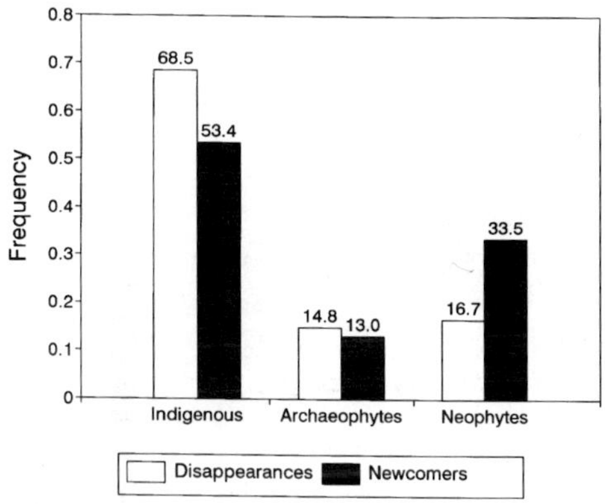

b

Fig. 1. Comparison with respect to the origin status of (a) the taxa present in species lists 1982 versus 1995, and of (b) the disappearances versus newcomers. Disappearances are those species recorded only in 1982, newcomers are those present only in 1995. Percentages for particular groups are given at the top of the bars.

Results

From 1982 to 1995, there was an increase in the total number of species by 26.3 % (taking the species number from 1982 as 100 %). This was the result of an increase in each of the categories distinguished with respect to the origin status (Table 1). Native species increased in their number by 21.4 % and archaeophytes by 10.2 %. However, the increase which was by far the most remarkable was that of neophytes (by 90 %). Consequently, the floristic spectrum with respect to the origin status changed in favour of neophytes (Fig. 1a) and this trend was statistically significant ($P<0.05$, Table 2).

If the set of species present only in 1980s and not recorded in 1995 (termed as "Disappearances") is compared with those not present in 1982 but recorded in 1995 (termed as "Newcomers"), the trend of the neophyte increase is also obvious (Table 1, Fig. 1b).

The comparison of neophyte species sets from both periods did not reveal any

Table 2. Differences in species lists (1982 versus 1995) classified according to species status (native versus archaeophytes or neophytes), and comparison of neophytes present in 1982 and 1985 from the viewpoint of origin, life forms and life strategies. The null hypotheses tested were: (1) Species present in the villages studied in 1982 do not differ in their status from those present in 1995; (2) The set of neophytes present in 1982 does not differ from that recorded in 1995 in origin, life form and life strategy. χ^2 test on contingency tables was used. Regions of origin distinguished are as in Fig. 2. Life forms distinguished: chamaephytes, geophytes, hemicryptophytes, phanerophytes, therophytes. Only the C-, CSR-, and CR-strategies (according to Grime 1979) were used for the test because of a low number of species possessing other strategies. NS = the null hypothesis cannot be rejected on significance level less than 0.05.

	df	χ^2-value	P
Species status	2	6.99	<0.05
Neophyte origin	4	1.79	NS
Neophyte life form	4	1.92	NS
Neophyte life strat.	2	0.29	NS

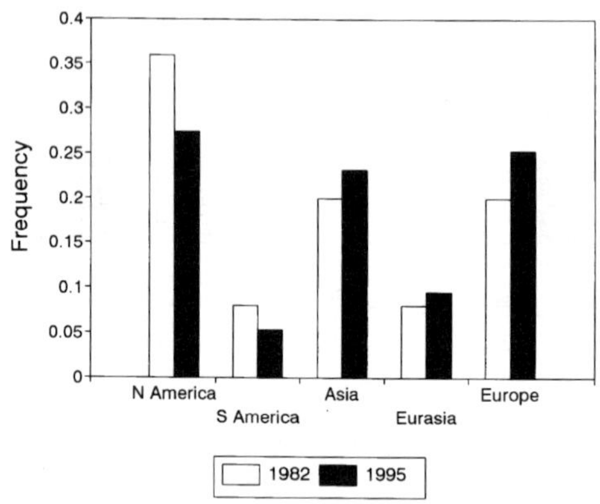

Fig. 2. Composition of the neophyte species sets according to the area of origin. Although there was a slight increase in representation of European and Asian species relative to those coming from North America, the difference between 1982 and 1995 was not significant (see Table 2).

significant differences in the frequency of life forms, life strategies or the area of origin (Table 2). In both periods, it was the North American species which were most strongly represented (Fig. 2).

The commoness of particular groups of species was analyzed on the basis of species occurrence in particular villages, i.e. their frequency (understood as the number of villages in which the species occurred). While the pattern has not actually changed for native species and archaeophytes (Fig. 3a, b), neophytes remarkably increased the frequency of their occurrence. In 1982, only 2.0 % of the total number of neophytes occurred in 9 or 10 villages whereas this is now 10.5 %. In the same vein, the percentage of neophytes occurring in at least 5 villages increased from 20.0 % to 36.8 % (Fig. 3c).

Fig. 4 indicates that those neophytes present in the villages studied in 1982 with high frequency had a better chance to be common in 1995. However, even among the newcomers, a principle group (35.2 %) became established in at least three villages. On the other hand, all the neophytes that disappeared between 1982 and

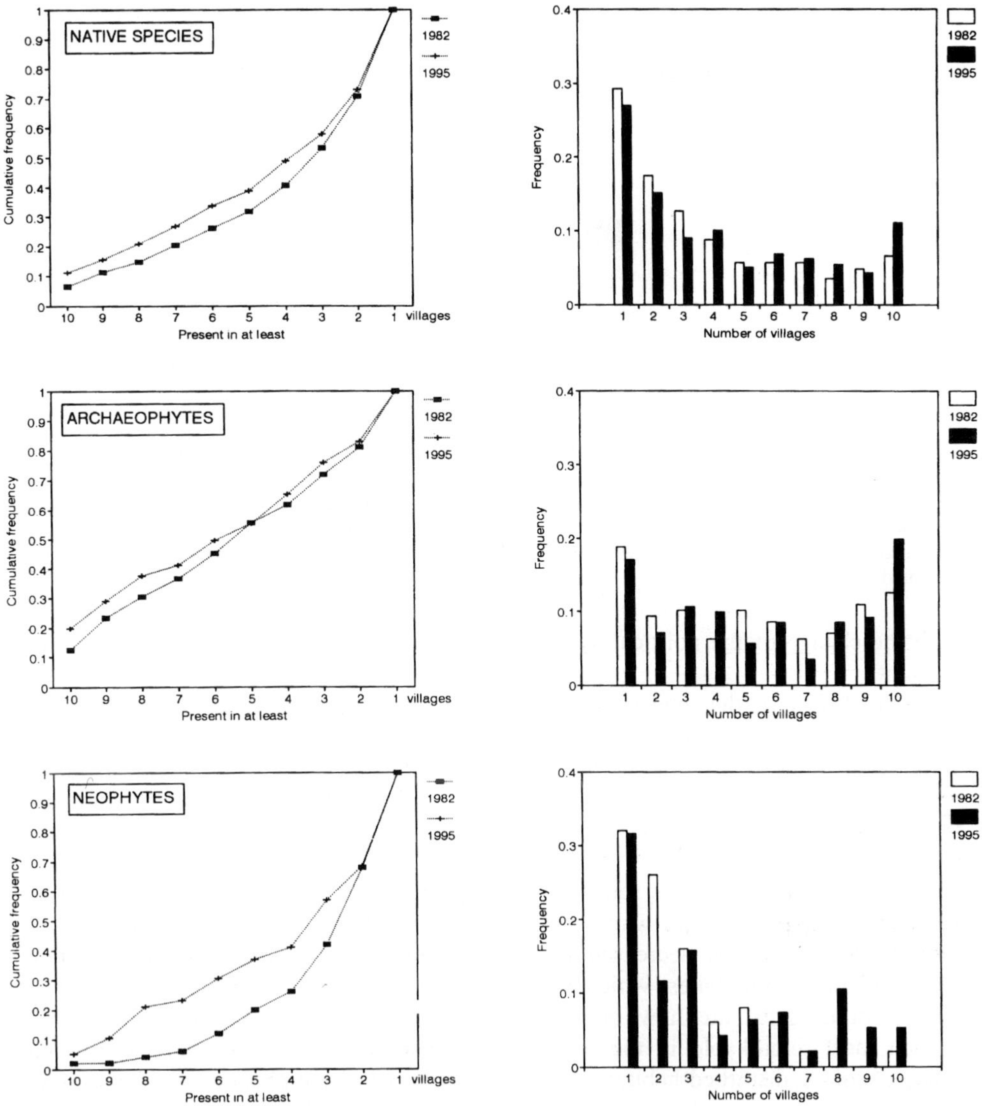

Fig. 3. Cumulative frequency of occurrences in the villages studied shown for (a) native species, (b) archaeophytes, and (c) neophytes. Frequency distribution of species with respect to the number of villages in which they occurred is shown in the left part of the diagram.

1995 were present with the lowest frequency (1–2 occurrences), or, put in another way, of those with the lowest frequency, 31.0 % disappeared.

Discussion

To explain the changes in the composition of village flora recorded in the present study, the changes in management and life style of Czech countryside (associated

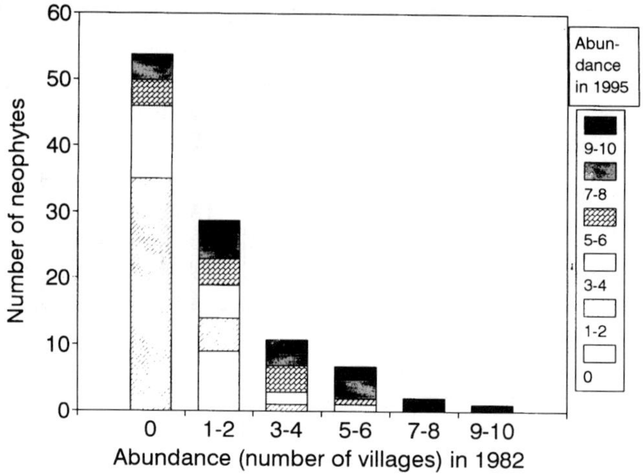

Fig. 4. Changes in abundance (expressed as the number of villages in which the species occurred) of neophytes from 1982 to 1995. Each bar representing the number of species with a given number of villages in which it occurred in 1982 is divided according to the number of villages these species occupied in 1995.

with the change of the political system at the end of the 1980s) must be taken into account.

In the 1980s, at the time of the first research period, there was no private agriculture, the arable land was managed by so-called cooperative farms and these farms were located mostly at the periphery of the village. The countryside was being overdosed with nutrients, especially nitrogen, coming from extensive fertilization – the average input of pure nutrients (N, P_2O_5, K_2O) was 255 kg per hectar per year (Moldan 1990).

At present, there is some return to private agriculture and the number of houses keeping cattle, poultry and waterfowl is increasing. Further, the input of fertilizers into the landscape has decreased remarkably and, as a consequence, quite a few weeds of oligotrophic sites, which used to be common in the past but diminished in the recent decades, are now present again. Also, the boom of building activity, though most remarkable in the big cities, is also obvious in the countryside and creates suitable disturbed habitats for ruderal flora. All these phenomena increase the diversity of habitats available in the villages and explain the increase in the species richness recorded by the present study.

But why was the increase in neophytes the most remarkable? The following reasons may serve as an explanation: (1) There is undoubtedly considerable increase in tourism as the area is one of the most attractive in the central Bohemian region. Although it is difficult to give an unequivocal evidence that a plant was brought in by tourism, more visitors coming undoubtedly mean more diaspores, often carried from very distant areas. (2) The variety of plant species and cultivars available on the gardening market is incomparably higher than 15 years ago. Numerous private companies were established in the last few years and the flow of germplasm into the country is practically uncontrollable. These plants, by escaping from cultivation, represent an important contribution to the neophyte pool, e.g. *Viola × wittrockiana, Sorbaria sorbifolia, Mahonia aquifolium, Papaver somniferum, Lychnis coronaria,*

Euphorbia lathyris, Alcea rosea and many others. (3) Neophytes, being mostly confined to man-made sites, are very much encouraged by building activity (Kowarik 1990), e.g. *Galinsoga parviflora, G. ciliata, Amaranthus powellii, Veronica persica* or *Conyza canadensis*, to mention those confined to disturbed areas.

The Bohemian Karts protected area is, by Czech standards, rather special area because of its natural beauty and nature conservation value (richness of flora and fauna). Its touristic attractivness, and proximity to Prague increase the possibilities of undesirable introductions of plant species. More data are therefore needed to allow broader generalization, but the present study undoubtedly indicates an increasing and disturbing importance of alien plant species in the contemporary Czech countryside. The increasing number of neophytes, the most dynamic and relatively recently introduced group of aliens, is especially alarming. The majority of them occur only temporarily, but on the other hand, some of the most noxious invaders are known to recruit from relatively recent introductions (Cronk and Fuller 1995).

Acknowledgments

Our thanks are to Max Wade and John Brock for their comments and for improving our English. The study was supported by the Grant Agency of the Czech Republic (grant no. 206/93/2440).

References

Ellenberg, H., Weber, H.E., Düll, R., Wirth, V., Werner, W. and Paulissen, D. 1991. Zeigerwerte von Pflanzen in Mitteleuropa. Scr. Geobot., Göttingen 18:1-284.

Cronk, Q.C.B. and Fuller, J.L. 1995. Plant invaders. Chapman and Hall, London.

Gilbert, O. 1989. Ecology of urban environment. Chapman and Hall, London.

Grime, J.P. 1979. Plant strategies and vegetation processes. John Wiley and Sons, Chichester.

Grime, J.P., Hodgson, J.G. and Hunt, R. 1988. Comparative plant ecology. Unwin and Hyman, London.

Hejný, S. and Slavík, B. (eds.) 1988-92. Flora of the Czech Republic. Vol. 1-3. Academia, Praha (in Czech).

Hobbs, R.J. 1989. The nature and effects of disturbance relative to invasions. In: Drake, J.A., Mooney, H.A., di Castri, F., Groves, R.H., Kruger, F.J., Rejmánek, M. and Williamson, M. (eds.). Biological invasions: A global perspective. pp. 389-405. John Wiley and Sons.

Holub, J. and Jirásek, V. 1967. Zur Vereinheitlichung der Terminologie in der Phytogeographie. Folia Geobot. Phytotax. 2:69-113.

Klotz, S. 1987. Floristische und vegetationskundliche Untersuchungen in Städten der DDR. Düsseldorf Geobot. Kolloq. 4:61-69.

Kowarik, I. 1990. Some responses of flora and vegetation to urbanization in Central Europe. In: Sukopp H., Hejný, S. and Kowarik, I. (eds.). Urban ecology. pp. 45-74. SPB Academic Publ., The Hague.

Kowarik, I. 1995. On the role of alien species in urban flora nad vegetation. In: Pyšek P., Prach K., Rejmánek M., and Wade P.M. (eds.). Plant invasions – general aspects and special problems. pp. 85-103. SPB Academic Publ., Amsterdam.

Kunick, W. 1974. Veränderungen von Flora und Vegetation einer Grossstadt, dargestellt am Beispiel von Berlin (West). Diss. Techn. Univ. Berlin.

Kunick, W. 1982. Zonierung des Stadtgebietes von Berlin (West). Ergebnisse floristischer Untersuchungen. Landschaftsentw. u. Umweltforsch., Berlin 14:1-164.

Moldan, B. 1990. Environment of the Czech Republic. Academia, Praha. [In Czech].

Pyšek, P. 1982. The vegetation of villages in the Bohemian Karst. Thesis, Charles Univ. Prague [in Czech].

Pyšek, P. 1985. Contribution to the village flora of the Bohemian Karst. Zpr. Čs. Bot. Společ., Praha 20:69-77. [in Czech]

Pyšek, P. 1991. Die Siedlungsvegetation des Böhmischen Karsts. 1. Syntaxonomie. Folia Geobot. Phytotax., Praha 26:225-262.

Pyšek, P. 1992. Die Siedlungsvegetation des Böhmischen Karts. 2. Ökologische Charakteristik. Folia Geobot. Phytotax., Praha 27:113-135.

Pyšek, P. 1993. Factors affecting the diversity of flora and vegetation in European human settlements. Vegetatio, Dordrecht 106:89-100.

Pyšek, P. 1995. On the terminology used in plant invasion studies. In: Pyšek P., Prach K., Rejmánek M. and Wade M. (eds.). Plant invasions - General aspects and special problems. pp. 71-81. SPB Academic Publ., Amsterdam.

Reidl, K. and Dettmar, J. 1993. Flora und Vegetation der Städte Ruhrgebietes, insbesondere der Stadt Essen und der Industrieflächen. Ber. Dt. Landeskunde 67:299-326.

Rejmánek, M. 1989. Invasibility of plant communities. In: Drake, J.A., Mooney, H.A., di Castri, F., Groves, R.H., Kruger, F.J., Rejmánek, M. and Williamson, M. (eds.), Biological invasions: a global perspective. pp. 369-388. John Wiley and Sons. Chichester.

Slavík, B. (ed.) 1995. Flora of the Czech Republic. Vol. 4. Academia, Praha (in Czech).

Sukopp, H. and Werner, P. 1983. Urban environment and vegetation. In: Holzner, W., Werger, M.J.A. and Ikusima, I. (eds.). Man's impact on vegetation. pp. 247-260. Dr. W. Junk Publ., The Hague.

Webb, D.A. 1985. What are the criteria for presuming native status? Watsonia 15:231-236.

Wittig R. 1991. Ökologie der Grossstadtflora. G. Fischer, Stuttgart.

THE EFFECT OF SUCCESSIONAL AGE AND DISTURBANCE ON THE ESTABLISHMENT OF ALIEN PLANTS IN MAN-MADE SITES: AN EXPERIMENTAL APPROACH

Marek Bastl[1], Petr Kočár[1], Karel Prach[1] and Petr Pyšek[2]
[1]*Faculty of Biological Sciences, University of South Bohemia, Branišovská 31, České Budějovice, CZ -370 05, Czech Republic;* [2]*Institute of Botany, Academy of Sciences of the Czech Republic, CZ - 252 43 Průhonice, Czech Republic*

Abstract

The hypothesis that younger successional stages are less resistant to invasions than older stages was verified experimentally in two kinds of primary successional seres (sand pit and peat-bog disturbed by peat extraction) in southern Bohemia, Czech Republic. Three comparable successional stages aged 0 (i.e. the year of disturbance), 10 and 25 years were selected in both sites. Eight species alien to Czech flora were sown into particular sites and stages and their establishment was recorded. MANOVA revealed significant ($P<0.001$) differences in the number of established seedlings between the successional stages and sites. The difference in seedling numbers among the three successional stages was significant (GLM, $P<0.05$) in seven species in the disturbed peat-bog. In the sand pit, significant differences were found in four species and the results from this site were more variable, showing no clear trend. In general, the 10 years old stages seemed most convenient for seedling establishment. The role of disturbance regime on seedling establishment was studied in the second experiment conducted in wet meadows near the town of Třeboň, southern Bohemia, Czech Republic. *Heracleum mantegazzianum* and *Impatiens glandulifera* were sown into plots varying in species composition, biomass and cover, and their germination was recorded. The correlation between the number of seedlings and any of the community characteristics measured was significant for neither species. The only trend observed was that the numbers of established *Impatiens glandulifera* seedlings decreased with increasing stand height. The results indicate that it is very difficult to find a community characteristic acting as a simple predictor of establishment success.

Introduction

Establishment of alien plants invading regions outside their native distribution range is a crucial step in the process of plant invasions (di Castri 1990). This stage of the process is rather poorly understood because of considerable lack of data, however it can be expected that the probability of successful seedling establishment would increase with the level of disturbance creating open space in a closed canopy (Crawley 1989). With respect to the successional status, young successional stages are predicted to be more susceptible to invasion by alien species than older stages (Rejmánek 1989; Hobbs and Huenneke 1992), and there is some support for this hypothesis coming from both theoretical models (Rejmánek 1989) and field data (Lepart and Debussche 1991). The latter, however, mostly compare *a posteriori* successional stages with respect to the representation of alien species (Kowarik

Plant Invasions: Studies from North America and Europe, pp. 191–201
edited by J.H. Brock, M. Wade, P. Pyšek and D. Green

1995). The major methodological flaw associated with such an approach is, however, that it is impossible to separate the effect of successional status from that of the input of diaspores which need not be constant over time (Crawley 1986). The later successional stages may then seem more resistant to invasions because, at least in part, the input of diaspores would decrease over time. To determine whether or not such an assumption is true, experimental data are needed.

The present paper explores the experimental verification of the hypothesis that younger successional stages are less resistant to invasions than older stages. Furthermore, the invasion success appears to be affected by such community characteristics as total cover and biomass (Rejmánek 1989; Hobbs 1989), i.e. those principally affected by disturbance regimes. There is some support for the hypothesis that disturbances in plant communities promote invasions (Hobbs and Humphries 1995). For that reason, the effects of various disturbance regimes on the seedling establishment during invasion are also addressed.

Study sites and methods

Effect of successional stage

The effect of successional age on the establishment of alien species seedlings was studied in two kinds of primary successional seres running in (1) sand pit land, (2) peat-bog disturbed by peat extraction. The localities were situated close to each other in southern Bohemia, Czech Republic (1-longitude 14°57'E, latitude 48°50'N; 2-longitude 14°48'E, latitude 48°57'N) at the altitude of 445-460 m a.s.l. Mean annual temperature was about 7°C, annual sum of precipitation was about 650 mm (50-year average).

Three comparable successional stages aged 0 (i.e. the year of disturbance), 10 and 25 years were selected in both sites. The age of particular successional stages was inferred from historical maps and site history records. The initial stages were completely free of vegetation. Ten years old stages in the sand pit were characterized by trees and shrubs up to 3 m in height (*Pinus sylvestris, Betula pendula, Salix caprea*) and by a sparse herbaceous layer. In the disturbed peat bog, ten year old stages were covered by scattered trees and shrubs (*Betula pendula, Pinus sylvestris, Rubus fruticosus* agg.), rather compact herb layer consisted mostly of *Molinia coerulea*. The oldest stages in both seres were covered by a dense pine-birch forest with low cover in the herbaceous layer.

A single measurement of ground water table was carried out in July 1994, using bore-holes installed in each successional stage. The sites and stages were similar from the viewpoint of moisture conditions. The position of ground water table slightly fluctuated around 1 m below surface.

Eight species considered as invasive in the Czech Republic (Pyšek *et al.* 1995) were used for experimental sowing (Table 1). The species were selected to cover a wide range of life forms and taxonomical groups. Seed availability and germination characteristics were also taken into account. In each site and successional stage, each species was sown into 4 randomly selected 1 × 1 m plots, giving the total of 24 plots per species. The seeds were sown in March 1995, before the onset of growing period, except for *Robinia pseudoacacia* and *Physocarpus opulifolius* which were sown later

Table 1. Species used in experiments on seedling establishment, their basic characteristics, and number of seeds sown in particular plots.

Species	Life form	Origin	Seeds per 1 m^2		
			Peat bog	Sand pit	Meadow
Acer negundo	tree	N America	200	200	–
Ailanthus altissima	tree	Asia/China	150	150	–
Bidens frondosa	annual	N America	400	400	–
Heracleum mantegazzianum	perennial	Asia/Caucasus	1000	1000	500
Lupinus polyphyllus	perennial	N America	45	90	–
Physocarpus opulifolius	shrub	N America	500	500	–
Pinus strobus	tree	N America	100	100	–
Robinia pseudoacacia	tree	N America	200	200	–
Impatiens glandulifera	annual	Asia/Himalayas	–	–	500

(May) to avoid possible frosts. *Robinia pseudoacacia* was treated before germination by hot water (90°C, 3 times for 15 seconds – Grime *et al.* 1981).

The number of seedlings were counted immediately after germination. Another census was carried out at the end of the vegetation period in October 1995. The data (seedling numbers) were analyzed by generalized linear models (GLM) (McCullagh and Nelder 1989) using the S-PLUS package (Anonymous 1995a, b) and after log-transformation also by MANOVA (Scheiner 1993).

Effect of community characteristics and disturbance regime

The experiment examining the role of disturbance regime on seedling establishment was conducted in a complex of wet meadows attached to a large fishpond, near the town of Třeboň, southern Bohemia, Czech Republic at the altitude of 430 m a.s.l.; latitude 48°61' N; longitude 14°55' E (see Jeník and Květ 1984; Prach 1993 for details on the site).

An homogenous, uncut sward dominated by sedges (*Carex gracilis, C. vesicaria, C. nigra, C. canescens*) and a grass *Calamagrostis canescens* was used as a study plot. Prior to the experiment reported in the present paper, the meadow was subjected (1984–1991) to various treatments (fertilization, mowing, trampling, burning) in a long-term experiment on response of species composition and productivity of a wet grassland to various kinds of disturbance (Prach and Soukupová, in press). As a result, the treated plots varied in species composition, biomass and cover, providing a convenient subject for testing the effect of these characteristics on the establishment of alien species.

Heracleum mantegazzianum and *Impatiens glandulifera* were sown in plots previously subjected to particular treatments at the beginning of January, the former in 1993 and the latter in 1994. In each plot, five subplots 1 × 1 m received 500 seeds of each species (Table 1). Seedlings and establishing plants were counted regularly.

Results

Effect of successional stage

Numbers of seedlings that became established and survived until the end of vegetation period are shown for particular sites and successional stages in Table 2.

Table 2. Establishment of seedlings and their survival until the end of the first vegetation period represented as the percentage of the total number of the seeds sown. Means from 4 plots are given in sand pit and disturbed peat bog, that from 5 subplots in the moist meadow (see Methods for details). The figures from the moist meadow experiment are based on the total number of seedlings germinated in 1994 and 1995, and are pooled over treatments.

Site	Sand pit			Peat bog			Meadow (pooled data)
Successional stage	0	10	25	0	10	25	
Acer negundo	0.9	0.9	11.0	0.8	3.0	0.1	–
Ailanthus alitissima	5.0	0.2	0.2	10.0	16.0	0	–
Bidens frondosa	0	0	0	0	0	0	–
Heracleum mantegazzianum	0	0	0	0	0	0	0
Lupinus polyphyllus	0	3.0	1.0	5.0	2.0	0	–
Physocarpus opulifolius	0	3.0	0	0.2	6.0	0	–
Pinus strobus	5.0	0	0	0	5.0	0	–
Robinia pseudoacacia	2.0	2.0	4.0	6.0	5.0	0	–
Impatiens glandulifera	–	–	–	–	–	–	0.4

MANOVA revealed significant ($P<0.001$) differences in the number of established seedlings among the successional stages and localities. The interaction between successional stage and site was significant at the same level, indicating that the differences between sites and successional stages have no additive character.

The difference in established seedling numbers among the three successional stages was significant (GLM, $P<0.05$) in seven species in the disturbed peat-bog. Generally, the 10 year old stages seemed as those most convenient for seedling establishment. In the sand pit, significant differences were found for four species and the results from this site were more variable, showing no clear trend (Fig. 1).

Effect of community characteristics and disturbance regime

Some seedlings of *Heracleum mantegazzianum* appeared in all but one treatment in 1993 and none survived into the next year. Another cohort germinated in spring 1995 and no seedling survived into the next year. *Impatiens glandulifera* germinated in the year when sown. In the heavily manured plot, 25.8 % of seedlings survived, the plants set fruit and a vital population established the following year (Table 3).

A similar germination response was found for both species, the number of *Heracleum* seedlings in particular plots being positively correlated with that of *Impatiens* ($r = 0.66$, $P = 0.05$). The highest seedling numbers of both species were recorded in the burned plot. The germination was, on the other hand, very low in trampled plots and especially in those subjected to moderate liming. No correlation between the

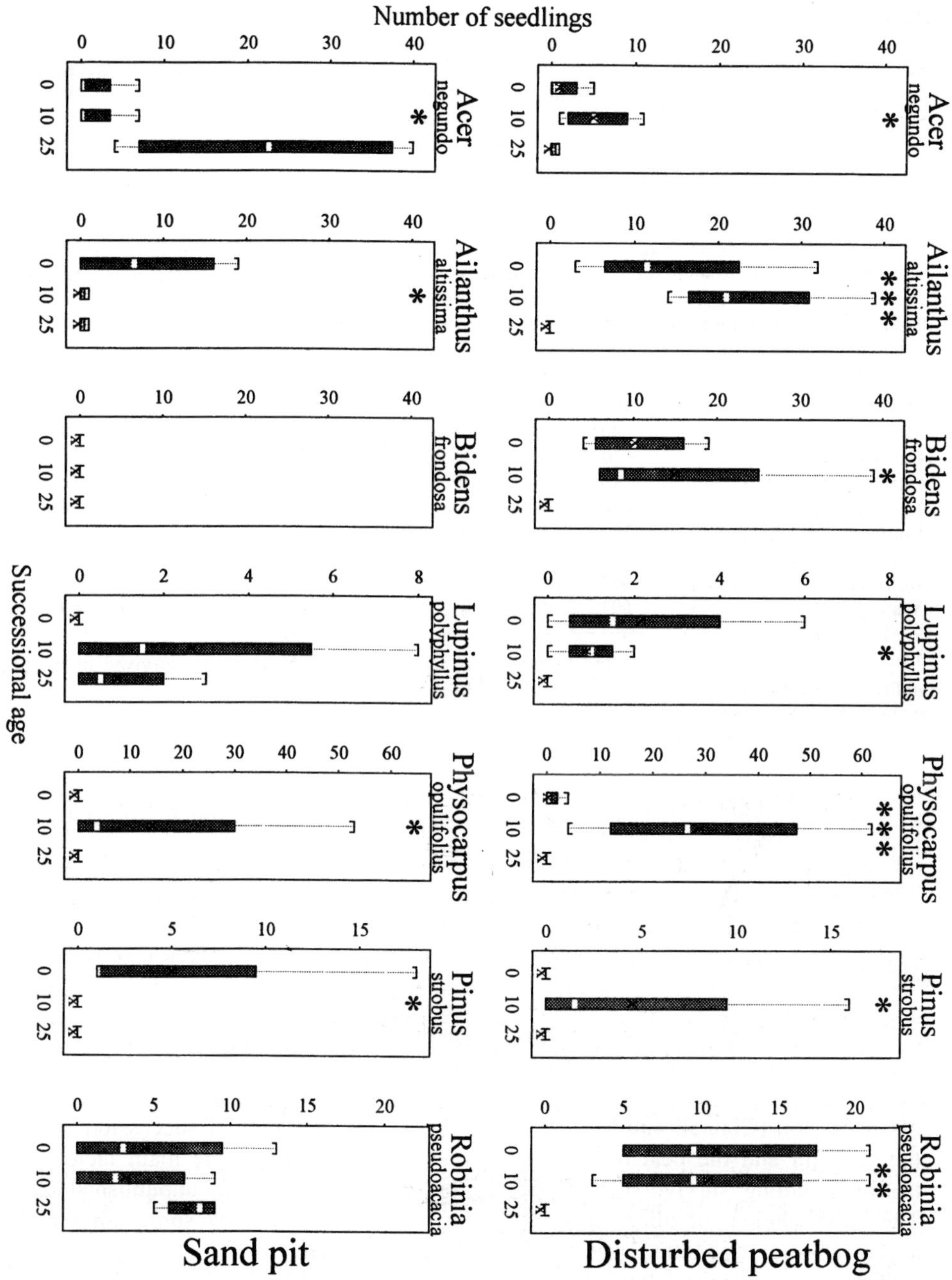

Fig. 1. Numbers of established seedlings (per 1 m^2) shown for particular sites (sand pit, disturbed peat bog) and successional stages (0, 10, and 25 years old). Note the different scale used to stress the comparison between sites and stages for each species. Significant differences in seedling numbers among the three successional stages as revealed by GLM are indicated (* $p<0.05$; ** $p<0.01$; *** $p<0.001$).

Table 3. Community characteristics and seedling establishment in the moist meadow experiment. Cover (%) was measured by the point-quadrat method, oven-dry biomass (48 hours at 85°C) is given in g.m^{-2}. Species diversity is expressed by Shannon index (log_2 base). Stand height is given in cm. Seedling numbers represent means from five 1 × 1 m subplots. No *Heracleum* seedlings were observed in 1994. No later recruitment of *Impatiens glandulifera* was observed except as indicated below.

Treatment	Community characteristics						Seedling establishment		
							Heracleum		Impatiens
	Living biomass	Litter	Dicott cover	Monocot cover	Stand height	Species diversity	1993	1995	1994
Control	277	400	47	88	80	2.87	11	11	49
Manuring intensive	276	300	35	98	95	2.57	10	16	31#
Manuring moderate	293	290	43	106	95	2.58	16	3	0
Cutting intensive	304	0	71	115	44	3.48	14	7	50
Cutting moderate	245	39	92	143	57	2.50	10	25	70
Liming intensive	494	440	58	147	80	2.95	17	22	57
Liming moderate	280	261	79	101	77	2.67	0	1	0
Liming, manuring	437	442	68	114	85	2.40	37	19	75*
Burning	287	62	77	127	51	2.97	53	109	100
Trampling	395	101	60	191	58	4.03	5	4	49

Eight plants flowered in 1994, and two of these set seed. In 1995, 41 plants survived in the plot until the reproductive period.
* Ten plants flowered, none of them set seed.

number of seedlings for either species for any of the measured community characteristics was significant. The only trend observed was that the numbers of established *Impatiens* seedlings tend to decrease with increasing stand height ($r = -0.51$, $P = 0.07$).

Discussion

Although the results vary with respect to particular species, a trend observed in this study was that mid-successional stages were the most favourable for the establishment of alien species. This pattern may be partly explained by the facilitation model of succession (Connell and Slatyer 1977). In the initial stages of succession in both the sand pit and the peat bog, topsoil moisture appears to be the main factor controlling seedling survivorship, especially in the summer (Bastl, unpublished data). The establishment of aliens might have also been restricted by some other rather severe abiotic conditions, such as the lack of nutrients.

Ten years after the colonization of the sites, moisture conditions tend to improve by the growth of native species. On the other hand, after the canopy becomes closed later in the course of succession, the lack of light, compact litter layer, and competition from the established herb layer are probably other factors inhibiting seedling performance in more advanced successional stages. The negative effect of litter on seedling establishment was repeatedly indicated in various experiments (Facelli and Pickett 1991; Sydes and Grime 1981a,b; Pyšek 1990).

The hypothesis predicting easier invasion of younger sites (Rejmánek 1989) does not seem to be fully supported by the present study. The pattern observed is more complicated and stresses the importance of timing in succession, i.e. that following a precocious arrival, establishment may be prevented because of the absence of resources whereas late entry may be prevented by established vegetation and natural enemies (Crawley 1989). Moreover, the available data largely do not take into account that the higher representation of invaders in early successional communities may be due to higher rates of species introductions. Hard data to separate this effect from the community resistance to invasion are not available, and there are indications of a contradicting pattern (Crawley 1986). Viewed from this perspective, our results may be taken as unbiased by the varying rate of seed introductions.

At a more detailed scale, represented by the wet meadow experiment, it appears very difficult to find a community characteristic that acts as a simple predictor of establishment success. The concept of "plant's eye view" and that of safe sites are difficult to record and quantify, still they play a crucial role in the process of seedling establishment (Harper 1977; Johnstone 1986). Strong disturbances, such as addition of large amounts of nutrients, probably enhance invasions (Hobbs and Huenneke 1992) as indicated by the heavily manured plot, the only one in the wet meadow experiment in which a population of *Impatiens glandulifera* was established. Nevertheless, the absence of clearcut results stresses the necessity to assess each particular invasion as a context-specific process (Lepart and Debussche 1991; D'Antonio 1993). The available experimental studies indicate that the effect of disturbance on the outcome of invasion depends on the invading species, invaded community, and the type of disturbance involved (Hobbs and Huenneke 1992; Knapp 1992).

The present study also documents that successful seedling establishment, a crucial step in the process of invasion, by far does not mean the start of successful invasion. In older successional stages and under a closed canopy, the further elimination of seedlings is probable due to competition for resources with established vegetation (Crawley 1989). Similarly, in the closed meadow sward the dense cover of resident species prevented the growth of seedlings folowing establishment and no seedling survived except in two experimental plots.

Field experiments in which seeds are introduced into relatively intact vegetation clearly indicates that for successful establishment of colonizers a large quantity of seeds must be introduced (Bazzaz 1986). For example, experimental introduction of seeds of *Melaleuca quinquenervia* and *Schinus terebinthifolius* (at least of 2 million of the former and 20,000 of the latter) both major invaders in Florida, into an array of communities representing various successional stages, degrees of disturbance, hydroperiods, fire regimes, and soil types yielded establishment rates between 0.01 and 2.62 percent (Myers 1983; Ewel 1986). Similarly in both species, introduced seeds yielded more seedlings in disturbed communities than in mature ones, and in all cases the number of surviving seedlings in mature communities was lower than that in man-made (Myers 1983; Ewel 1986). However, in many communities seed introductions were complete failures and yielded no surviving seedlings (Ewel 1986). Despite careful design of these experiments in order to exclude the possibility that timing of the seed-introduction experiments did not coincide with the chance juxtaposition of conditions required to ensure both germination and seedling establishment, the establishment rates obtained were still very low and confirm that repeated

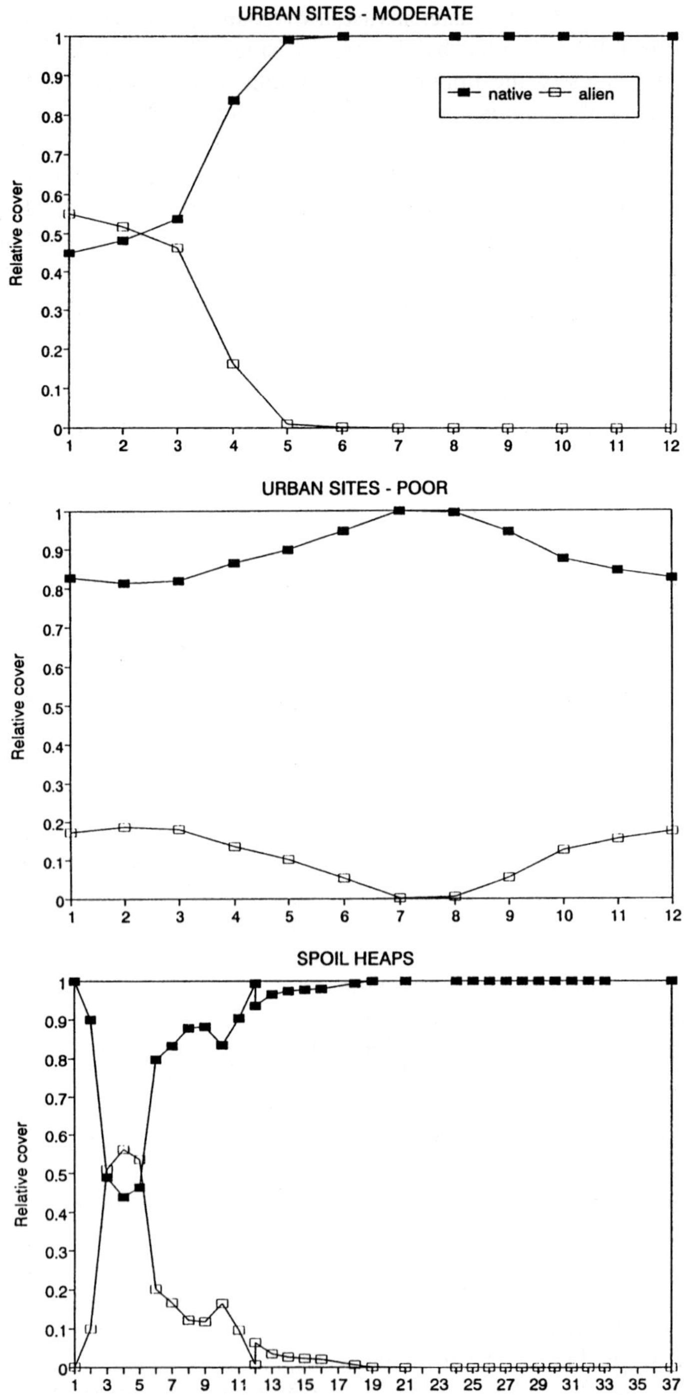

a.

Fig. 2. Participation of alien species in successional seres of man-made habitats: urban sites – nutrient poor; urban sites – moderate in nutrients; and spoil heaps from coal mining. All the seres were located in the western part of the Czech Republic, and are described in details in Prach *et al.*(1993), Prach and Pyšek (1994a,b). Only the seres with significant representation of alien species are →

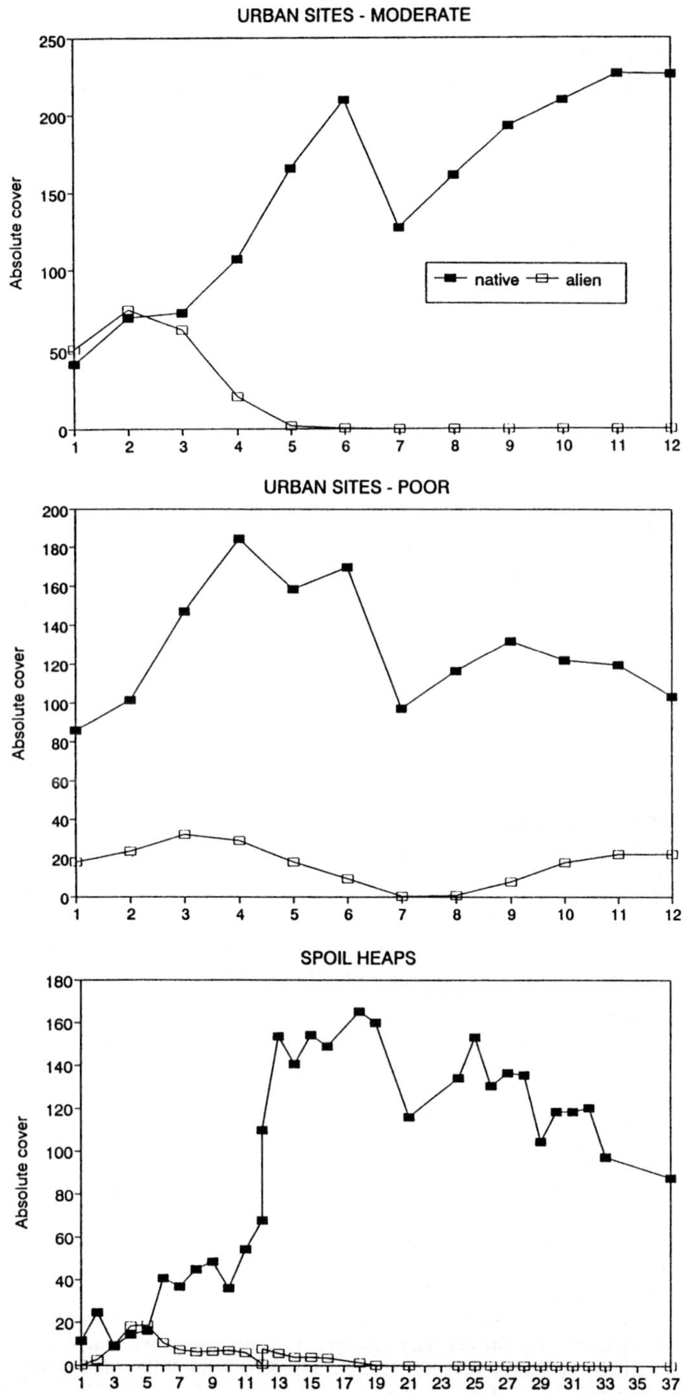

b.

shown. (a) relative cover, (b) absolute cover. Alien species were also present in reclaimed plots in the areas deforested due to air pollution, in nutrient rich urban sites, wet spoil heaps and xeric abandoned fields. However, in these habitats, their cover did not exceed 5 %. Aliens were absent from the rest of the sites (sand pit, peat bog, emergent bottom, mesic and wet abandoned fields).

introductions of a large number of disseminules may be necessary for initial establishment of colonizers (Martins and Jain 1979).

The barriers alien species have to cope with during particular phases of invasion are reflected in representation of aliens in a wide array of successional seres in man-made habitats of central Europe. Among the 15 seres analysed (Prach *et al.* 1993; Prach and Pyšek 1994a,b), the alien species only play an important part in urban areas (Fig. 2), i.e. in sites with sufficiently high and permanent input of diaspores (Kowarik 1995), and partly in spoil heaps from coal mining. The presence of aliens in drier and nutrient poor sites such as spoil heaps may be from reduced competitive ability of native species in these less favourable habitats.

Acknowledgments

Our thanks are due to Veronika Kostečková and Rudolf Hlaváček for field assistance. The study was supported by grants of the Grant Agency of the Czech Republic nos. 204/93/2440 and 204/94/0395.

References

Anonymous. 1995a. S-PLUS guide to statistical and mathematical analysis, Version 3.3 for Windows. StatSci, a division of Mathsoft, Inc., Seattle, Washington.

Anonymous. 1995b. S-PLUS user's manual, Version 3.3 for Windows. StatSci, a division of Mathsoft, Inc., Seattle, Washington.

Bazzaz, F.A. 1986. Life history of colonizing plants: some demographic, genetic, and physiological features. In: Mooney, H.A. and Drake, J.A. (eds.): Ecology of biological invasions of North America and Hawai, pp. 96-110, Springer Verlag, New York.

Connell, J.M. and Slatyer, R.O. 1977. Mechanisms of succession in natural communities and their role in community stability and organization. American Naturalist 111: 1119-1144.

Crawley, M.J. 1986. Plant ecology. Blackwell Sci. Publ., Oxford.

Crawley, M.J. 1989. Chance and timing in biological invasions. In: Drake, J.A., Mooney, H.A., di Castri, F., Groves, R.H., Kruger, F.J., Rejmánek, M. and Williamson, M. (eds.) Biological invasions: A global perspective, pp. 407-423, John Wiley and Sons, Chichester.

D'Antonio, C.M. 1993. Mechanisms controlling invasion of coastal plant communities by the alien succulent *Carpobrotus edulis*. Ecology 74: 83-95.

di Castri, F. 1990. On invading species and invaded ecosystems: the interplay of historical chance and biological necessity. In: di Castri, F., Hansen, A.J. and Debussche, M. (eds.) Biological invasions in Europe and the Mediterranean Basin. pp. 3-16, Kluwer Academic Publ., Dordrecht.

Ewell, J.J. 1986. Invasibility: Lessons from South Florida. In: Mooney, H.A. and Drake, J.A. (eds.): Ecology of biological invasions of North America and Hawai, pp. 214-230, Springer Verlag, New York.

Facelli, J. M. and Pickett, S. T. A. 1991. Plant litter: its dynamics and effects on plant community structure. Bot. Rev. 57: 1-32.

Grime, J.P., Mason, G., Curtis, A.V., Rodman, J., Band, S.R. and Mowforth, M.A. 1981. A comparative study of germination characteristics in a local flora. J. Ecol. 69: 1017-1059.

Harper, J.L. 1977. Population biology of plants. Academic Press, London.

Hobbs, R.J. 1989. The nature and effects of disturbance relative to invasions. In: Drake, J.A., Mooney, H.A., di Castri, F., Groves, R.H., Kruger, F.J., Rejmánek, M. and Williamson, M. (eds.) Biological invasions: A global perspective. pp. 389-405, John Wiley and Sons, Chichester.

Hobbs, R.J. and Huenneke, L.F. 1992. Disturbance, diversity and invasion: implication for conservation. Conserv. Biol. 6: 324-337.

Hobbs, R.J. and Humphries, S.E. 1995. An integrated approach to the ecology and management of plant invasions. Conserv. Biol. 9: 761-770.

Jeník, J. and Květ, J. 1984. Long-term research in the Třeboň Biosphere Reserve, Czechoslovakia.

In: di Castri, F., Baker, F.W.G. and Hadley, M. (eds.) Ecology in practice, Part 1: Ecosystem management. pp. 437-459, Tycooly, Unesco, Paris.

Johnstone, I.M. 1986. Plant invasion windows: a time based classification of invasion potential. Biol. Rev. 61: 369-374.

Knapp, P.A. 1992. Secondary plant succession and vegetation recovery in two western Great Basin Desert ghost towns. Biol. Conserv. 60: 81-89.

Kowarik, I. 1995. On the role of alien species in urban flora and vegetation. In: Pyšek, P., Prach, K., Rejmánek, M. and Wade, M. (eds.) Plant invasions – general aspects and special problems. pp. 85-103, SPB Academic Publ., Amsterdam.

Lepart, J. and Debussche, M. 1991. Invasion processes as related to succession and disturbance. In: Groves, R.H. and di Castri, F. (eds.) Biogeography of Mediterranean invasions. pp. 159-177. Cambridge Univ. Press, Cambridge.

Martins, P.S. and Jain, S.K. 1979. The role of genetic variation in the colonizing ability of rose clover (*Trifolium hirtum* All.). Amer. Natur. 113: 591-595.

McCullagh, P. and Nelder, J.A. 1989. Generalized linear models. Ed. 2. Chapman and Hall, London.

Myers, R.L. 1983. Site susceptibility to invasion by the exotic tree *Melaleuca quinquenervia* in southern Florida. J. Appl. Ecol. 20: 645-658.

Prach, K. 1993. Vegetational changes in a wet meadow complex, South Bohemia, Czech Republic. Folia Geobot. Phytotax. 28: 1-13.

Prach, K. and Pyšek, P. 1994a. Spontaneous establishment of woody plants in Central European derelict sites and their potential for reclamation. Restor. Ecol. 2: 190-197.

Prach, K. and Pyšek, P. 1994b. Clonal plants – what is their role in succession? Folia Geobot. Phytotax. 29: 307-320.

Prach, K. and Soukupová, L. (in press). Vegetation changes in the wet meadows under anthropogenic impact. In: Květ, J. and Jeník, J. (eds.). Ecological basis for sustainable management of wetlands in the Třeboň Biosphere Reserve. Unesco, Paris.

Prach, K., Pyšek, P. and Šmilauer, P. 1993. On the rate of succession. Oikos 66: 343-346.

Pyšek, P. 1990. The influence of *Calamagrostis villosa* on the species diversity of deforested sites in the Krušné Hory Mts. Preslia 62: 323-335.

Pyšek, P., Prach, K. and Šmilauer, P. 1995. Relating invasion success to plant traits: an analysis of the Czech alien flora. In: Pyšek, P., Prach, K., Rejmánek, M., Wade, M. (eds.) Plant invasions – general aspects and special problems. pp. 39-60, SPB Academic Publ., Amsterdam.

Rejmánek, M. 1989. Invasibility of plant communities. In: Drake, J.A., Mooney, H.A., di Castri, F., Groves, R.H., Kruger, F.J., Rejmánek, M. and Williamson, M. (eds.) Biological invasions: A global perspective. pp. 369-388, John Wiley and Sons, Chichester.

Scheiner, S.M. 1993. MANOVA: Multiple response variables and multispecies interactions. In: Scheiner, S.M. and Gurevitch, J. (eds.) Design and analysis of ecological experiments, pp. 94-112, Chapman and Hall, London.

Sydes, C. and Grime, J.P. 1981a. Effects of tree leaf litter on herbaceous vegetation in deciduous woodland. I. Field investigations. J. Ecol. 69: 23-248.

Sydes, C. and Grime, J.P. 1981b. Efects of tree leaf litter on herbaceous vegetation in deciduous woodland. II. An experimental investigations. J. Ecol. 69: 249-262.

SEASONAL DYNAMICS OF AERIAL BIOMASS OF *FALLOPIA JAPONICA*

Petr Horn
Faculty of Biological Sciences, Branišovská 31, České Budějovice, CZ – 37005, Czech Republic

Abstract

The dynamics of aerial biomass of *Fallopia japonica* Houtt. was studied during the vegetation season 1995 in the Czech Republic, Central Europe. The results show rapid growth of standing crop in the beginning of season together with small increase of aerial biomass caused by flowering later in August and September.

Introduction

Fallopia japonica Houtt. is native in Japan, Northern China and Taiwan. This alien species is widely distributed in the British Isles, and much of the Central and northern Europe (Grime *et al.* 1988). *F. japonica* first appeared in the Czech Republic in the end of the 19th century as ornamental plant (Pyšek and Prach 1993).

Fallopia japonica occurs mainly on habitats strongly influenced by man, although it is spread commonly along river banks. It is characteristic to relatively nutrient-rich habitats and it is a fast growing competitive species (Beerling *et al.* 1994).

R. japonica is an exceedingly tall herb (even more than 2.5 m), and a dense canopy develops each summer beneath only few species are capable to survive. Coexistence of other species with *F. japonica* is restricted further by the persistent litter which accumulates within established stands. The shoots die back each autumn, and early growth of new shoots in the spring is supported by reserves in the thick fleshy rhizomes. All these factors allow *F. japonica* to form virtual monocultures (Grime *et al.* 1988).

The question addressed in this study is whether aerial biomass or its dynamics can explain the high competitive ability of the species.

Methods

The study site was situated in the area of railway station at the town of České Budějovice, South Bohemia, elevation at 380m, Czech Republic, Central Europe. The site was sufficiently wet and fertile and never mown during research period.

Material was sampled in seven dates: May 3rd, May 12th, June 3rd, June 16th, July 10th, August 17th, October 25th 1995. At least twenty shoots were randomly

Plant Invasions: Studies from North America and Europe, pp. 203–206
edited by J.H. Brock, M. Wade, P. Pyšek and D. Green

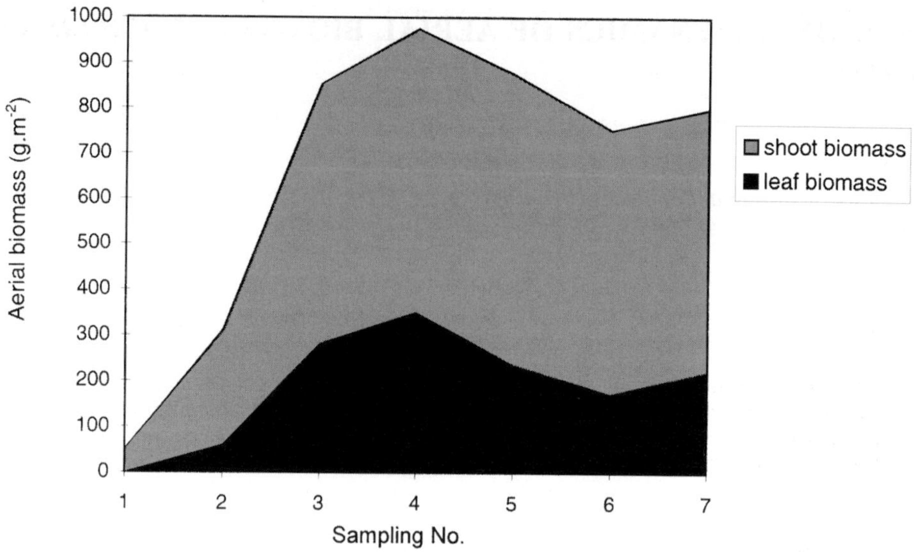

Fig. 1. Dynamics of aerial biomass of *Fallopia japonica.*

selected at each date and cut at the stem base, shoots and leaves were separated. After measuring stem diameter at 5 cm above earth surface, shoots were oven dried for 24 hours at 90°C and weighted.

At the same time two transects consisting of 5 plots 0.5 × 0.5 m each were established from the border to the center of the stand. Diameter of all rooting shoots was measured 5 cm above earth surface.

Allometric equations were calculated for the relationship between basal stem diameter and biomass (Table 1). Biomass was estimated using a regression model (Pielou 1974)

Results

Maximum total aerial biomass was estimated at 976 $g.m^{-2}$ on June 16. At the same time maximum leaf biomass attained 351 $g.m^{-2}$. The results are presented in Table 1 and Figure 1.

Discussion

The total aboveground biomass corresponds well to the values published by Brock (1995), Callaghan *et al.* (1984) and Beerling (1990) from United Kingdom. Another estimation (Horn and Prach 1994) from the Czech Republic is also consistent with this study. The small difference is caused by stand conditions together with different time of sampling.

Standing crop of *F. japonica* increases twice, first very rapidly in May, being supported by stored nutrients from previous season, with a second growth peak

Table 1. Seasonal changes in total, shoot and leaf biomass in 1995.

Date (1995)	total biomass ($g.m^{-2}$)	R^2	leaf biomass ($g.m^{-2}$)	R^2	shoot biomass ($g.m^{-2}$)	R^2
May 3rd	49	0.91	0		49	0.91
May 12th	320	0.82	59	0.53	253	0.87
June 3rd	887	0.89	281	0.80	573	0.93
June 16th	976	0.93	351	0.89	624	0.93
July 10th	896	0.91	235	0.69	641	0.93
August 10th	762	0.80	171	0.53	581	0.86
September 25th	869	0.92	220	0.66	579	0.94

R^2 (coefficient of determination) the proportion of total variation in biomass that is explained or accounted for by the fitted regression
regression equation: log (biomass) = a+ b × log (basal stem diameter)

observed in September during the flowering period, when increase of biomass is probably caused by new branching.

The increase of stem length in May is rapid. A growth rate given by Wolf (1971) from USA was 4.65 cm per day. *F. japonica* profits from this fast growth early in the season (Grime *et al.* 1988) when it quickly overtops surrounding vegetation, attaining a competitive advantage. On the other hand, relative growth rate (RGR) of *F. japonica* in this time is not especially high – $R_{max} = 0.7\ g.g^{-1}.week^{-1}$, so using only RGR, *F. japonica* would not be a very successful competitor (Beerling 1990).

If we compare biomass of *F. japonica* with Czech native species, which often grow in the same habitat, we argue that standing crop itself is not probably responsible for strong competitive ability of this alien species (Horn and Prach 1994).

Standing crop can partly support vegetative spread of *F. japonica* along river banks, because there is evidence of regenerating new plants from cut stems from greenhouse experiments (Brock and Wade 1992).

The role of root competition and litter in excluding other herbs is still unclear. Underground biomass was estimated by Brock (1994) at near 1500 $g.m^{-2}$. This is even greater than aerial biomass. Beside high underground biomass *F. japonica* has capacity for rapid above- and under-ground spread. Also, litter is relatively high, in December there was still 530 $g.m^{-2}$ of persisting material, which can inhibit establishment of other species (Horn, unpublished data).

Acknowledgments

I thank my supervisor K. Prach for his comments and the Faculty of Biological Sciences for financial support. The study was also partly supported by the grant of the Grant Agency of the Czech Republic no. 600393.

References

Beerling, D.J. 1990. The ecology and control of Japanese knotweed on river banks in South Wales. University of Wales, School of Pure and Applied Ecology, Ms. Thesis.

Beerling, D.J., Bailey, J.P. and Conolly, A.P. 1994. *Fallopia japonica* (Houtt.) Ronse Decraene. Biological flora of the British Isles, Journal of Ecology 82: 959-979.

Brock, J.H. and Wade, P.M. 1992. Regeneration of Japanese Knotweed from rhizomes and stems: observations from green house trials. In: Proceedings of the 9th International Symposium of the Biology of Weeds, pp.85-94, Dijon, France.

Brock, J.H. 1995. Technical note: Standing crop of *Reynoutria japonica* in the autumm of 1991 in the United Kingdom, Preslia 66: 337-343.

Callaghan, T.V., Scott, R., Lawson, G.J. and Manwaring, A.M. 1984. An experimental assessment of native and naturalised species of plants as renewable sources of energy: III. Japanese Knotweed – *Reynoutria japonica*. Institute of Terrestrial Ecology, Cumbria.

Grime, J.P., Hodgson, J. and Hunt, R. 1988. Comparative plant ecology. A functional approach to common British species. Unwin Hyman, London.

Horn, P. and Prach, K. 1994. Aerial biomass of *Reynoutria japonica* and its comparison with that of native species. Preslia 66: 345-348.

Pielou, E.C. 1974. Population and community ecology, pp.117-121, Gordon and Breach Science Publishers, New York.

Pyšek, P. and Prach, K. 1993. Plant invasion and the role of riparian habitats: a comparison of four species alien to central Europe. Journal of Biogeography 20: 413-420.

Wolf, F. 1971. The growth rate of *Polygonum cuspidatum*. Journal of Tennessee Academy of Science 46: 80.

THE USE OF GIS IN THE MANAGEMENT OF *FALLOPIA JAPONICA* IN THE URBAN ENVIRONMENT

Lois Child[1] and Louise de Waal[2]
[1]International Centre of Landscape Ecology, Department of Geography, Loughborough University, Loughborough, Leicestershire, LE11 3TU, UK;
[2]Division of Environmental Science, School of Applied Sciences, University of Wolverhampton, Wulfruna Street, Wolverhampton, WV1 1SB, UK

Abstract

Fallopia japonica (Japanese knotweed) in the City of Swansea, Wales, UK, is used as a case study to demonstrate the important contribution which Geographical Information Systems (GIS) can make to the integrated management of invasive plant species in the urban environment. The ecology of the species, its distribution, mode of regeneration and dispersal are briefly described. A comprehensive distribution survey of the plant in the City of Swansea was carried out in order to assess the extent of the invasion and survey data were captured onto Swansea City Council's GIS. Assessment of the survey data has revealed that the area covered by *F. japonica* in the City totals 48 ha which represents 0.5% of the total land area of the City. Major habitats invaded by *F. japonica* are shown to be sites subject to disturbance or sites receiving little or no management such as railway embankments, road verges, river and streambanks, cemeteries and waste industrial ground. Using the GIS database, correlations are possible between the survey data and land use and planning policy. The results of these correlations are being used by the City Council to underpin ongoing management of the plant.

Introduction

The idea of large scale computerised geographical mapping is not new. The Canadian Land Inventory CSIS (Canada Soils and Information System) was pioneered in 1965 and has been followed by a number of national land resource systems in the USA and UK (Briggs and Mounsey 1989). Developments in the sphere of information technology are occurring rapidly and Geographical Information Systems (GIS) are becoming increasingly important as a management tool in land use planning, integrating environmental and other data and allowing monitoring and evaluation. This is particularly true when dealing with extensive areas with complex land use and ownership issues such as those faced by local authorities. Recent studies in Canada which have used inventory data include the assessment of the depletion of ecological resources from forest fires; establishing moose and caribou habitat management programmes; formulating recreation planning; targeting watershed management; evaluating hazardous waste management; and assisting in waterway clean up programmes (Heywood 1990).

The case study presented in this paper deals with the integrated management of the garden escape *Fallopia japonica* (Japanese knotweed) in the City of Swansea,

Plant Invasions: Studies from North America and Europe, pp. 207–220
edited by J.H. Brock, M. Wade, P. Pyšek and D. Green

All Local Plan policies	
Constraints:	tree preservation orders (ongoing)
	listed buildings
	ancient monuments
	national nature reserves
	local nature reserves
	Sites of Special Scientific Interest
	ancient woodlands
	conservation areas
Hazards:	landfill sites, landfill buffers
	notified hazards
Policy areas:	planning brief
	Article 4 Directions (special planning control power which enables the City Council to protect certain areas)
Environment:	Japanese knotweed data
	habitat data (ongoing)
	Source: Weston, 1995

Fig. 1. Strategic planning information entered onto Swansea City Council's GIS.

South West Wales, UK. It represents a good example of best practice by a local authority in terms of allocation of available resources, integration between council departments and co-operation between the various bodies dealing with a forward planning attitude towards sustainable management within the City. In 1991–92, Swansea City Council implemented a GIS for use primarily for development control and forward planning (Weston 1995). The City Council has prepared a strategic list of information valuable in the planning process which has been entered onto the GIS (Fig. 1).

The City of Swansea, with a population of approximately 190,000, has a history of industrial development mainly associated with the coal mining and steel industry dating from the industrial revolution at the turn of the 20th century. In recent years, industrial decline has led to the closure of mines and other heavy industry leaving behind a legacy of many hectares of disturbed and often derelict land. For many years the spread of the invasive plant *F. japonica* within the City Council boundary was known to be considerable, much of it associated with disturbed and derelict land but no accurate assessment had been made. In order to assess the extent of the problem a comprehensive distribution survey of the plant was carried out in 1992, with the aim of entering the survey data onto the recently acquired GIS. The survey data, once captured would provide a baseline for monitoring future control programmes and further spread of the plant, would aid the City Council in co-ordinating control efforts and allow integration of the survey data with other data sets held on the system.

Geographical Information Systems

Geographical Information Systems (GIS) are best described as an integrated collection of hardware, software, data and liveware which operates in an institutional context. A GIS is characterised by a great diversity of applications and is therefore difficult to capture within a single definition. Several of the many definitions applied to GIS are quoted by Maguire (1991) and include the following:

- A powerful set of tools for collecting, storing, retrieving at will, transforming and displaying spatial data from the real world.
- A special case of information systems where the database consists of observations on spatially distributed features, activities, or events, which are definable in space as points, lines, or areas. A GIS manipulates data about these points, lines, and areas to retrieve data for *ad hoc* queries and analyses.

The first definition highlights one of the four approaches in GIS applications, the toolbox approach, emphasising the generic aspects of GIS. The second definition signifies the database approach in GIS as an important and powerful tool. A third approach is process-oriented, stressing the information handling capabilities of GIS and a fourth is the application approach which looks at GIS as a problem addresser. The importance of GIS as a decision support and management information system is also widely recognised. Hence, it is not surprising that the use of GIS has become more common in local authorities since the mid 1980s and is still growing. One of the main reasons why the use of GIS has become popular on a worldwide basis and especially within central and local government is because GIS is a special case of information systems with three main foci: the map, the database and the ability to perform spatial analysis. The map focuses on the cartographic aspects of map processing and displaying, the database signifies the importance of a well-designed and implemented database and the spatial analysis focuses on the analysis and modelling properties of GIS.

The first official involvement of the UK government in handling geographical information and related issues dates back to the formation of the Ordnance Survey of Great Britain. This led initially to the development of a single national series of maps based on a countrywide National Grid and later to the production of digital maps (Chorley and Buxton 1991). Since almost all of the work and responsibilities of local authorities involves some kind of spatial data an active interest has been taken in the application of GIS in recent years (Fig. 2). Plymouth City Council implemented a GIS in 1989 (Markham *et al.* 1990) and around the same time Berkshire County Council started their GIS pilot study, initially to assist in highway maintenance and strategic land use planning (Lodwick and Cushnie 1990). Prior to the use of GIS, the analysis of spatial and temporal data was a time consuming and laborious process. Advantages of using GIS, once the initial data have been digitised onto the system, include ease of regular updates and analysis using the overlay and buffering facilities (Goodchild 1991). In this way, spatial and temporal variability of data can be taken into account.

The use of GIS in the urban environment is wide ranging and includes applications such as industrial location modelling, supply and demand forecasting in real estate markets, utility applications, land suitability and facility siting (Parrott and Stutz 1991). Although these socio-economic applications in the urban setting may be more significant in historical terms, a multitude of environmental data and applications have become increasingly important and are being integrated into the GISs of local authorities. These data sets can include environmental information on topics such as land use, habitat, designated areas, emissions into the air, and water and soil quality. Swansea City Council implemented their GIS primarily for development control and forward planning, however four departments, Estates, Leisure Services, Engineers and Planning, have access to the GIS and each is adding data including all those relating to local plan policies (Fig. 1). The Estates Department has a long-term project

Monitoring changes in resources (land and building, equipment and infrastructure) and conditions (economic, social, demographic, environmental, etc.)
Forecasting changes in housing requirements, school roles, travel patterns, the economy and demand for land, leisure and community services.
Service planning through identifying and forecasting changes in patterns of need for services and investments as a basis for the delivery of services and deployment of resources. This will determine both the scale of provision and its location, it will also highlight areas of social deprivation.
Resource management of, for example, building maintenance, refuse collection, grass cutting, route scheduling of supply vehicles, mobile libraries, social service ambulances.
Transport network management including provision and maintenance of highways, public transport, school transport, street cleaning.
Public protection and security systems including police command and control systems, definition of police beats, location of fire hydrants, patterns of crime and incidents of fire.
Property development and investment including the preparation of development plans, assessing land potential and preparing property registers, promoting industrial development and rural resource management.
Education use of wide range of data for teaching purposes, including the use of demonstration software packages.

Source: Chorley and Buxton, 1991

Fig. 2. UK local government responsibilities involving spatial information.

to transfer the land terrier onto GIS. This register of land area and ownership has been extended to include an asset register of all City Council-owned properties in Swansea. Data from the 1991 Population Census have also been entered onto the GIS and the Planning Department is now entering habitat and nature conservation data, which includes the *F. japonica* survey data from this case study. A similar study carried out in Germany in an non-urban environment monitored submerged macrophytes. The use of GIS enabled the creation of geographically referenced maps of distribution areas of single species or species groups and analysis of changes in area and locations of species during the investigation period (Schmieder 1995). It was envisaged that similar results could be achieved by the application of GIS in the distribution survey of an invasive plant, *F. japonica*, in an urban environment. One of the most important features of the use of GIS in this kind of application is that many different data sets can be accessed simultaneously and manipulated (Parrott and Stutz 1991).

Fallopia japonica (Japanese knotweed)

In order to carry out a successful management programme for an invasive plant species it is essential to have a good understanding of the ecology of the plant. A brief description of *F. japonica* follows, for a more detailed description of the species and its control see Beerling *et al.* (1994) and Child *et al.* (1992).

Fallopia japonica Ronse Decraene (syn. *Reynoutria japonica* Houtt., syn. *Polygonum cuspidatum* Sieb. and Zucc.) (Japanese knotweed) is native to Japan and was introduced into many countries including Europe (Beerling *et al.* 1994) and the north-eastern United States (Pridham and Bing 1975) as an ornamental garden plant in the late 19th century. Since its introduction, it has become an increasing problem

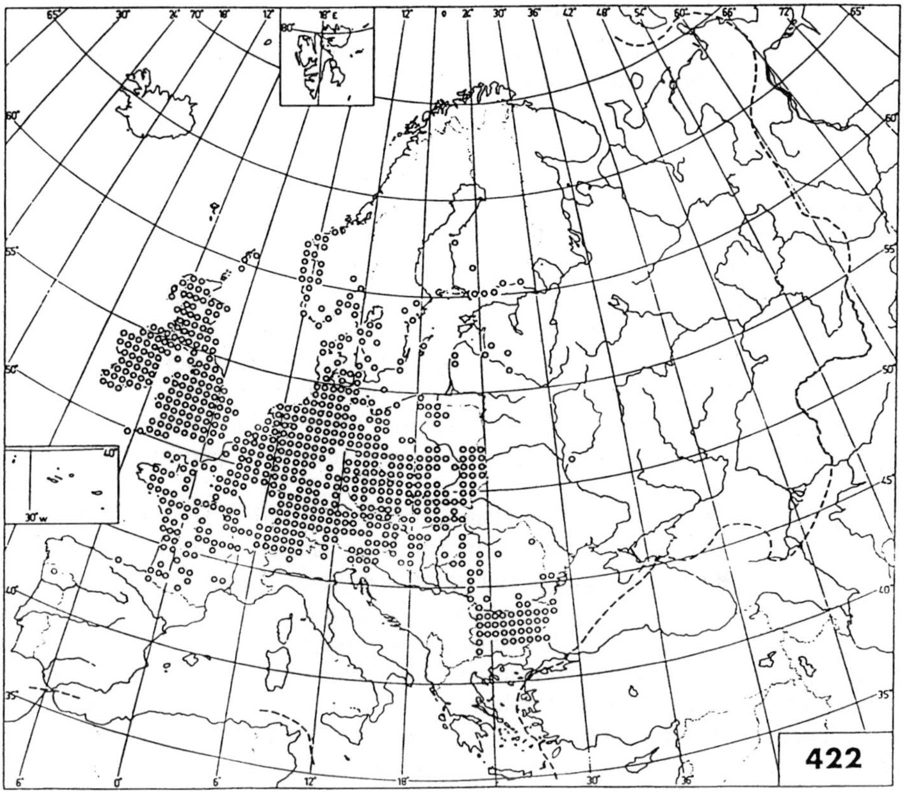

Fig. 3. Distribution map of *Fallopia japonica* in Europe (Jalas and Suominen, 1979)

in a variety of habitats, showing a typical exponential invasion pattern (Pyšek and Prach 1993). *F. japonica* is now widespread in Europe (Fig. 3) and in the north-eastern United States (Locandro 1973; Pridham and Bing 1975). The first naturalised escape in the British Isles was recorded in 1886 on cinder tips at Maesteg in South Wales (Conolly 1977) with the plant now recorded in approximately half of the 10 km squares used to map species distribution in the British Isles (Beerling *et al.* 1994).

The plant is not restricted to urban land, neither is it an agricultural weed, but tends to be associated with disturbed land, in particular, land associated with human disturbance. It is recorded in a wide range of habitats including along road verges and railway embankments, in parks, cemeteries and on waste ground in rural and urban areas and along water courses (Child *et al.* 1992).

F. japonica grows vigorously from an extensive rhizome system in early spring, with a growth rate of up to 46.5 mm in a single day (Wolf 1971). It forms dense stands which can reach a height of 2–3 m in one growing season. The plants die back in winter to leave thickets of distinctive orange/brown coloured erect, hollow, semi-woody stems which may persist for several years, eventually breaking down to form a deep litter.

Whereas the plant in its native habitat is dioecious, only the female plant has so far been recorded in the British Isles (Bailey *et al.* 1995). The means of dispersal of *F. japonica* in the British Isles is therefore vegetative, primarily by rhizome fragments but also through sections of cut stem material. The rhizome system is extensive and has been recorded as spreading up to 7 m away from a parent plant (Richards, Moorehead and Laing 1990) with estimates of underground biomass in the region of 16,000 kg ha^{-1} in monoculture stands (Brock 1995). Recent studies have shown that the regeneration potential of the rhizome material is considerable, with fragments of rhizome weighing 0.7 g sufficient to give rise to a new plant (Brock and Wade 1992). Regeneration from fresh stem material has also been highly successful both in water and in a soil medium (Brock *et al.* 1995). Natural dispersal of the plant along rivers and streams is by means of fragments of rhizome and stem material broken off the parent plant and washed downstream by high water flows. Human activities such as the cartage of earth contaminated with fragments of rhizome and transportation of stems and rhizomes in garden waste are important in the long distance dispersal of the plant and the invasion of new sites particularly in urban areas.

The problems associated with the plant are many and varied, not least in riparian areas where the extensive rhizome system can cause damage to flood defence structures (Beerling 1991). The dead aerial stems can cause instream blockage when washed into watercourses under high winter flows causing problems with flooding especially in urban areas and the dense growth of the plant restricts access to riverbanks for recreation and bank inspection. Furthermore, the dense canopy restricts growth of ground flora and reduces native species diversity in both rural and urban habitats. Particular problems in the urban environment include invasion of amenity areas and considerable damage to paving, car parking areas and even buildings.

Although the plant has been shown to be susceptible to chemical treatments using herbicides such as glyphosate, 2,4 D-amine, picloram and triclopyr (de Waal 1995; Beerling 1990; Scott and Marrs 1984), the use of some herbicides is restricted in riparian areas. The cost of application and the time necessary for an effective control programme (at least 3 years) often makes chemical treatment prohibitively expensive.

In view of the scale of the problem in Swansea, it is both unrealistic in practical and cost terms and undesirable in environmental terms (due to the large quantities of herbicide that would be required) to treat all existing *F. japonica* in the City effectively. It is therefore necessary to formulate a management strategy which minimises expenditure on blanket treatments and instead uses resources available to target key sites and prevent further spread.

Survey methodology

The distribution survey of *F. japonica* in the City of Swansea was carried out by the International Centre of Landscape Ecology during the early spring and late autumn of 1992. The plant is easily recognisable at these times due to the distinctive height and orange/brown colour of the dead woody stems and the presence of very little other vegetation. An area of the City totalling approximately 96 km^2 was systematically surveyed by a team of student volunteers equipped with 1:2,500 scale Ordnance

Survey (OS) maps and specially designed recording sheets (Fig. 4). Prior to the survey the volunteers had been trained in recognition of the plant and use of the recording techniques. The distribution of the species was recorded directly onto both the maps and the recording sheets. Each occurrence of the plant was drawn onto the OS maps, either as a cross for small areas (<40 m^2) or drawn onto the map as the actual extent of the stand for larger areas. Mountain bikes allowed increased coverage of all roads, pathways and tracks in the City with each pair of volunteers able to cover an area of between 1 and 3 km^2 in a day depending on the number of occurrences of the plant. The recording sheets were used not only for the purpose of achieving consistency in the field, but also to accomplish easy coding for data entry onto the GIS. Each recording sheet contained a variety of information including a code number for each entry, the site grid reference, the exact area covered by the plant (in m^2), the associated land use and the proximity of each site to water. Furthermore, detailed information on the height and management of each *F. japonica* stand was noted.

The accuracy of data obtained by means of visual inspection, particularly with regard to the position and area covered by plants could have been improved by the use of Global Positioning Systems (GPS).

Transfer of survey data onto GIS

The survey information was entered onto a computer system using Genamap, a GIS produced by Genasys II Ltd. Two levels of data were utilised, area data and attribute data. The area data, taken from the 1:2,500 OS field maps, outline the extent of the distribution of the plant within each entry. These data were entered onto new data layers by on-screen digitising, using digital OS base maps as reference layers. On-screen digitising requires an human operator to move a cursor over the screen thus capturing the positions of points and areas, building up digital representations of point, line, and area features. The attribute data taken from the recording sheets were entered onto separate files within the associated database of Genamap. The attribute data files contain all the relevant data for each separate area of *F. japonica* surveyed. Each entry made on the recording sheet was coded appropriately and subsequently entered onto the database of the GIS within a specified field. Once both the area and attribute data were entered, the information was stored under a specifically created work-area 'Jap_knotweed' within the GIS Genamap. This work-area or directory contains all the *F. japonica* maps and attribute data. Future updates may be stored as separate files within this specified work-area.

Once the attribute data were entered and tagged to correspond to the digitised area data, it was possible to view each stand of *F. japonica* on screen. This information could be correlated with other data stored on the GIS. The attribute data could also be manipulated as an independent database to give summary information for any field recorded on the survey recording sheets.

Japanese Knotweed Survey Recording Sheet

International Centre of Landscape Ecology, Department of Geography, Loughborough University, Loughborough, LE11 3TU. Tel: 01509 223030 Fax: 01509 223931

Recorded by	Date
Site name	
Grid ref.	Site ref.

Number of stands (**Cut**)		
Total area of cut stands	m	m
Number of stands (**Uncut**)		
Total area of uncut stands	m	m

N.B. Mark location on the map, as + when stand is <20x20m or linear feature is <25m long, otherwise map as total area.

Average height of stand		<1m		1-2.5m		>2.5m		
Max. stem diameter at 30cm above ground		<1cm		1-2cm		2-4cm		>4cm
Vegetation composition		Japanese knotweed only				mixture of Japanese knotweed and other vegetation		

Proximity to water courses		yes		no		
Slope		flat		moderate		steep

Land Use

☐ Housing	☐ Shops	☐ Business/Industrial	☐ Public buildings
☐ Garden	☐ Park	☐ Landscaped area	☐ Recreation grounds
☐ Farm land	☐ Woodland	☐ Graveyard	☐ Waste ground
☐ Car park	☐ Road verge	☐ Roundabout	☐ Railway embankment
☐ River bank	☐ Stream side	☐ Canal	☐ Dock
☐ Pond	☐ Sea front	☐ Other, specify	

Remarks: ....

Fig. 4. Recording sheet for the *Fallopia japonica* distribution survey.

Results and discussion

Assessment of the survey database shows that the total coverage of *F. japonica* in Swansea City is 48 ha which represents 0.5% of the total land area of the City. Approximately 60% of the coverage occurs on City Council owned land. However, the distribution of the plant throughout the City is not even, with the number of occurrences of the plant in a one kilometre square ranging from absent to over 70 locations. When correlating the presence of the species with land use, the survey shows that *F. japonica* occurs most frequently on land which has little or no management i.e. road verges, railway embankments, cemeteries, wasteground, stream and riverbanks, industrial land and amenity areas (Fig. 5). The information is not only useful in identifying the localities of the plant for future control efforts, but also in identifying the land owners and agencies involved in the future management of the species.

The City Council has identified three priority areas with regard to the control efforts for *F. japonica* (Hill 1994). These are: sites of high nature conservation value; areas identified for environmental improvement and landscaping as part of the Council's "Greening the City" initiative and sites where the infestation is causing damage or is creating a hazard.

The GIS survey data are now being used to refine these priority areas allowing identification of sources of potential new infestations e.g. watercourses where rapid spread of the plant to new sites could be controlled. Further areas are being identified where funding opportunities could be used to include *F. japonica* treatment, e.g. sites funded under the Urban Programme and sites where resources are available for long term maintenance. Additionally, areas can now be highlighted where the plant on City Council owned land is causing problems for other people e.g. spreading into private gardens and areas which are currently "Japanese knotweed free" where prevention of infestation will be a priority. Until recently the City Council was unable to monitor the spread of the plant as there was no baseline distribution data for this invasive species. Since the completion of the distribution survey and the instigation of the GIS, the City Council is able to monitor the plant and treatment methods can be targetted accordingly. Treatment is carried out either by the City Council's Direct Labour Organisation or by outside contractors. The GIS database provides an excellent tool with which to record the programme of treatment methods and to monitor effectiveness through future surveys.

New developments have in the past been responsible for increasing the distribution of the plant in the City, often by transportation of rhizome fragments in soil. Using GIS to correlate any planned land use change or future development with the *F. japonica* survey data, allows the City Council to put precautionary measures into place before any works begin. This is especially important in schemes such as highway improvements where large quantities of soil are moved with a high risk of spreading the plant. The GIS facilitates linking of different data layers, for example, combining the layers "Japanese knotweed data" and "Planning Brief" (Fig. 1) allows the City Council to identify whether or not a site is infested with *F. japonica* prior to a proposed new development. Figure 6 shows output from the GIS for a section of the City with these links in place. The proposed highway improvement scheme intersects an area of Japanese knotweed indicating an increased risk of causing the plant to spread by movement of earth at this point. With regard to such developments by outside contractors, the City Council now incorporates detailed specifications on the control of *F. japonica* into their contract documents (Welsh Development Agency 1994). These measures also ensure that where possible, treatment costs will be incorporated into the costs of redevelopment. The City Council is now taking precautionary measures to ensure that the plant is not caused to spread at infested sites or

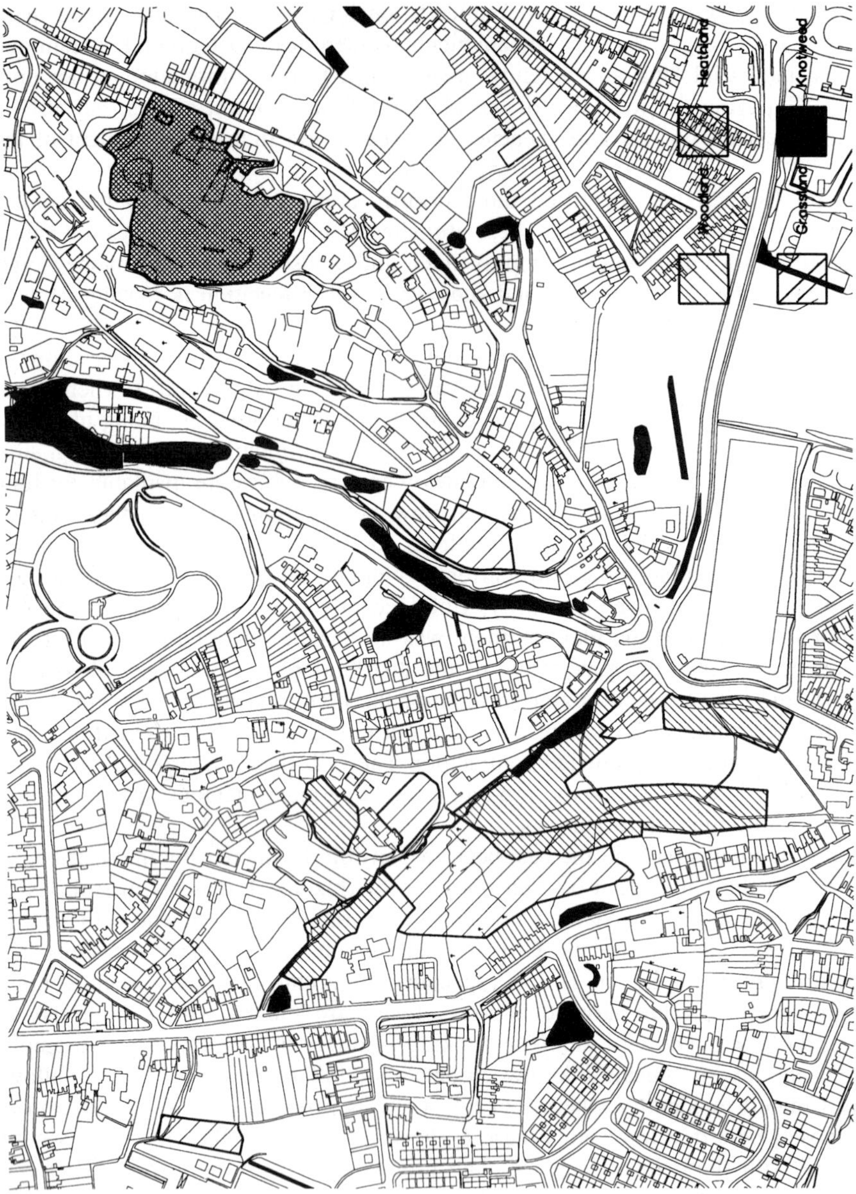

Fig. 5. Output from the GIS showing the presence of *Fallopia japonica* (dark shaded areas) in Swansea City. Permission to reproduce see Fig. 6.

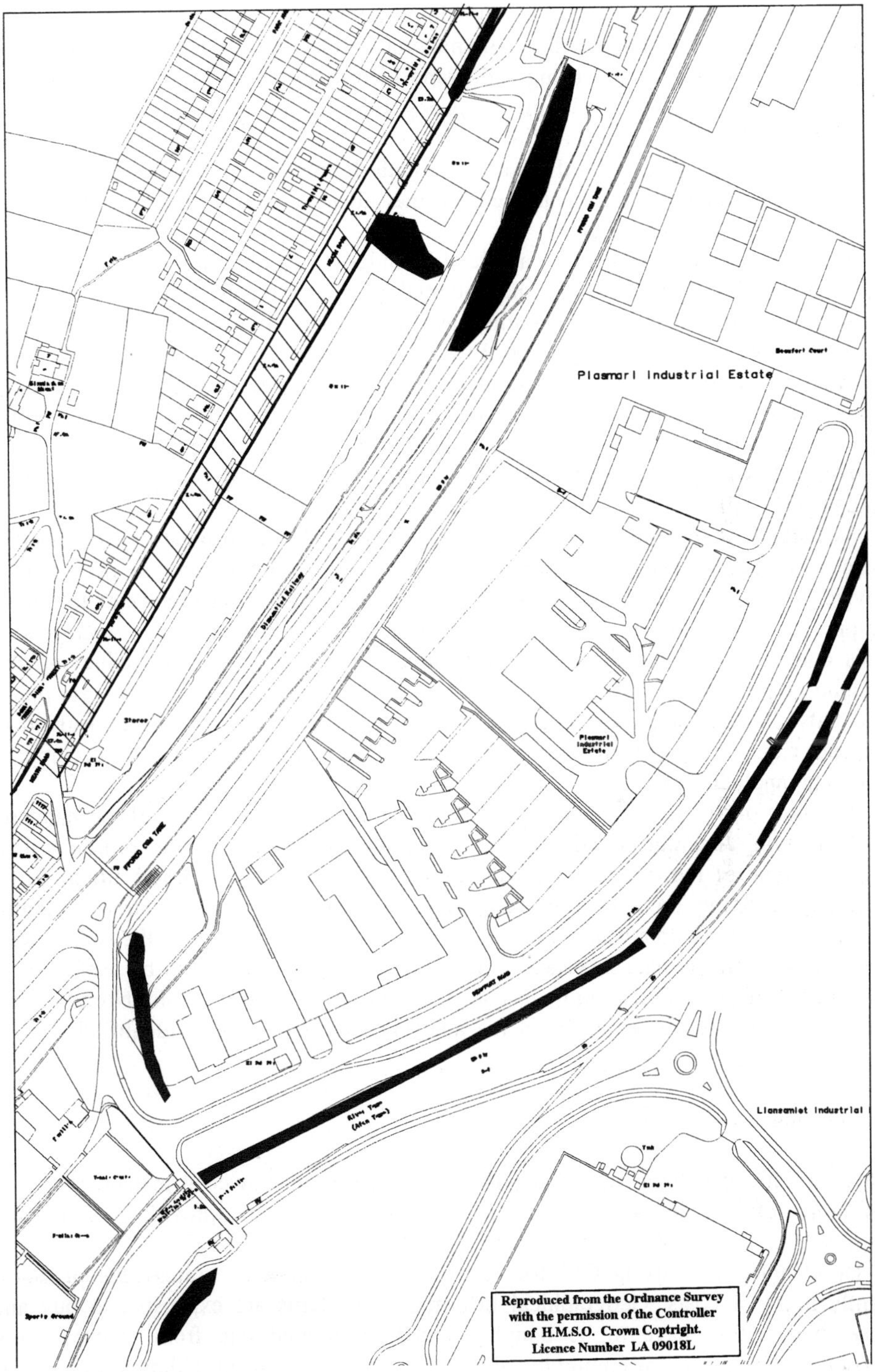

Fig. 6. Output from the GIS showing correlation between the presence of *Fallopia japonica* (dark shaded areas) and proposed highway improvements (hatched area) in Swansea City.

brought onto non-infested sites through importation of soil contaminated with rhizome fragments. To achieve this, soil material loads are inspected on delivery, or material is inspected *in situ* using a rhizome identification key designed specifically for *F. japonica* (Welsh Development Agency 1994). Soil which is found to be contaminated with *F. japonica* material is required to be either treated appropriately on site or disposed of at a licensed landfill site with the recommendation that it is buried at least 5 m deep. These measures ensure that all areas subject to development are protected from further invasion.

As plants are not restricted by artificial boundaries, an important aspect of control has to be achieved by removing the potential of infestation from adjacent sites and especially from sites upstream in the case of riparian environments. This may be realised through co-ordinating control efforts with adjacent landowners (Hill 1994; de Waal *et. al.* 1995). The GIS has identified that Swansea City Council is a major landowner of infested areas. Further analysis of the data allows identification of other major landowners, such as British Rail and The Environment Agency in addition to private landowners. In order to co-ordinate the management of *F. japonica*, all these landowners should be involved in future control efforts.

As part of its integrated strategy, the City Council has also identified the need to raise awareness of the problem and ways of dealing with it through provision of advice, information and training. This would be targetted at all sectors of the population, council employees, developers, contractors, landowners, land managers of sites containing the plant and the general public. Possible ways of achieving this include preparation of information sheets and procedural notes for the general public and City Council staff, arranging training sessions for contractors and staff from relevant Council departments and preparing guidelines for developers to accompany planning applications.

A review of the *F. japonica* survey methodology highlights the need for standardised surveying techniques and well trained surveyors. The use of the recording sheet reduced some of the inaccuracies which may have otherwise occurred although, for example, deciding on a priority landuse in a situation such as a stream running through an amenity area was not always a simple matter. The aims of the survey were achieved and all the data recorded on the sheets have been found to be useful in the assessment of the problem and subsequent interpretation and analysis. No additional data that may have been useful in the analysis have so far been identified.

Conclusion

This case study demonstrates that GIS is a valuable and powerful management tool for more rational decision making and increased efficiency, providing an holistic approach to the management of an invasive plant.

Disadvantages of using GIS for this type of management strategy are predominantly those of costs. Geographical Information Systems are expensive to buy and base maps either have to be bought, rented or created e.g. from colour photogrammetry. The costs of employing surveyors is also significant, although these costs can be kept to a minimum if volunteers are used. Digitising the data onto digitised base maps is a time consuming process but once the data are entered, the costs of running the system are small, especially if the system is required for a multipurpose use as in the Swansea example.

The advantages of GIS stored data are significant especially when the unique ability for spatial searching and overlaying are taken into account. The data in this case study have become available to a wide range of departments throughout the local authority, manipulation of the data and correlation between local planning policy and invasion alert areas is now available at the touch of a button and the problems of updating and storing paper survey sheets have been removed. GIS provides a robust system of data storage which is responsive to changes in planning policy, land use and land ownership well into the future.

Acknowledgements

The authors acknowledge funding from Swansea City Council for the survey and data entry and would like to thank Deborah Hill and John Weston for their support. Thanks also to National Rivers Authority (now The Environment Agency), Welsh Region for funding the survey of the River Tawe.

References

Bailey, J.P., Child, L.E. and Wade, P.M. 1995. Assessment of the genetic variation and spread of British population of *Fallopia japonica* and its hybrid *Fallopia* x *bohemica*. In: P. Pyšek, K. Prach. M. Rejmánek and M. Wade (eds.) Plant Invasions – General aspects and special problems. pp. 141-150. SPB Academic Publishing, Amsterdam.

Beerling, D.J. 1990. Short communication: the use of non-persistent herbicides, glyphosate and 2,4-D amine, to control riparian stands of Japanese knotweed (*Reynoutria japonica* Houtt.). Regulated Rivers: Research and Management 5: 413-417.

Beerling, D.J. 1991. Technical note: The testing of cellular concrete revetment blocks resistant to growths of *Reynoutria japonica* Houtt. (Japanese knotweed). Water Research 25: 495- 498.

Beerling, D.J., Bailey, J.P. and Conolly, A.P. 1994. Biological Flora of the British Isles *Fallopia japonica* (Houtt.) Ronse Decraene. Journal of Ecology 82: 959-979.

Briggs, D. and Mounsey, H. 1989. Integrating land resource data into a European geographical information system: practicalities & problems. Applied Geography 9: 5-20.

Brock, J.H. 1995. Technical note: Standing crop of *Reynoutria japonica* in the autumn of 1991 in the United Kingdom. Preslia 66: 337-343.

Brock, J.H. and Wade, P.M. 1992. Regeneration of Japanese knotweed (*Fallopia japonica*) from rhizome and stems: Observations from greenhouse trials. IXe Colloque International sur la Biologie des Mauvaises Herbes, 85-94. Dijon, France.

Brock, J.H., Child, L.E., de Waal, L.C. and Wade, P.M. 1995. The invasive nature of *Fallopia japonica* is enhanced by vegetative regeneration from stem tissues. In: P. Pyšek, K. Prach. M. Rejmánek and M. Wade (eds.) Plant invasions – General aspects and special problems. pp. 131-140. SPB Academic Publishing, Amsterdam.

Child, L.E., de Waal, L.C., Wade, P.M. and Palmer, J.P. 1992. Control and management of *Reynoutria* species (knotweed). Aspects of Applied Biology 29: 295-307.

Chorley, R. and Buxton, R. 1991. The government setting of GIS in the United Kingdom. In: D.J. Maguire, M.F. Goodchild and D.W. Rhind, (eds.) Geographical Information Systems: principles and applications. pp. 67-79. Longman Scientific and Technical, Harlow.

Conolly, A.P. 1977. The distribution and history in the British Isles of some alien species of *Polygonum* and *Reynoutria*. Watsonia 11: 291-311.

Goodchild, M.F. 1991. The technological setting of GIS. In: D.J. Maguire, M.F. Goodchild and D.W. Rhind (eds.) Geographical Information Systems: principles and applications. pp. 45-54. Longman Scientific and Technical, Harlow.

de Waal, L.C. 1995. Treatment of *Fallopia japonica* near water – a case study. In: P. Pyšek, K. Prach. M. Rejmánek and M. Wade (eds.) Plant invasions – General aspects and special problems. pp. 203-212. SPB Academic Publishing, Amsterdam.

de Waal, L.C., Child, L.E. and Wade, P.M. 1995. The management of three alien invasive riparian plants: *Impatiens glandulifera* (Himalayan balsam), *Heracleum mantegazzianum* (giant hogweed) and *Fallopia japonica* (Japanese knotweed) In: D. Harper and A. Ferguson (eds.) The ecological basis of river management. pp. 315-321. Wiley, Chichester.

Heywood, I. 1990. Monitoring for change. A Canadian perspective on the environmental role for GIS. Mapping Awareness 4: 24-26.

Hill, D.J. 1994. A practical strategy for the control of *Fallopia japonica* (Japanese knotweed) in Swansea and the surrounding area, Wales. In: L.C. de Waal, L.E. Child, P.M. Wade and J.H. Brock (eds.) Ecology and management of invasive riverside plants. pp. 195-198. Wiley, Chichester.

Jalas, J. and Suominen, J. 1979. Atlas Florae Europaeae. 4. *Polygonaceae*. Helsinki.

Locandro, R.R. 1973. Reproduction ecology of *Polygonum cuspidatum*. PhD thesis Rutgers University, The State University of New Jersey.

Lodwick, A. and Cushnie, J. 1990. A GIS pilot study in Berkshire. Mapping Awareness 4: 39-42.

Maguire, D.J. 1991. An overview and definition of GIS. In: D.J. Maguire, M.F. Goodchild and D.W. Rhind (eds.) Geographical Information Systems: principles and applications. pp. 9-20. Longman Scientific and Technical, Harlow.

Markham, R., Rix, D. and Bentley, R. 1990. GIS at Plymouth City Council – Coming to terms with reality. Mapping Awareness 4: 10-13.

Parrott, R. and Stutz, F.P. 1991. Urban GIS applications. In: D.J. Maguire, M.F. Goodchild and D.W. Rhind (eds.) Geographical Information Systems: principles and applications. pp. 247-260. Longman Scientific and Technical, Harlow.

Pridham, A.M.S. and Bing, A. 1975. Japanese bamboo. Plants and Gardens 31: 56-57.

Pyšek, P. and Prach, K. 1993. Plant invasions and the role of riparian habitats: a comparison of four species alien to Central Europe. Journal of Biogeography 20: 413-420.

Richards, Moorehead and Laing Ltd. 1990. Japanese knotweed (*Reynoutria japonica*) in Wales. Welsh Development Agency, Cardiff.

Schmieder, K. 1995. Application of Geographical Information Systems (GIS) in lake monitoring with submersed macrophytes at Lake Constance – Conception and purposes. Acta Botanica Gallica 142: 551-554.

Scott, R. and Marrs, R.H. 1984. Impact of Japanese knotweed and methods of control. Aspects of Applied Biology 5: 291-296.

Welsh Development Agency 1994. Model tender document for the eradication of Japanese knotweed. Welsh Development Agency, Cardiff.

Weston, J. 1995. Planning is paramount: GIS in district councils. Mapping Awareness 10: 32-35.

Wolf, F.T. 1971. The growth rate of *Polygonum cuspidatum*. Journal of the Tennessee Academy of Science 46: 80.

INDEX

Only those species are included which are treated in detail.

PLANT INVASIONS: STUDIES FROM NORTH AMERICA AND EUROPE